BIOCHEMISTRY RESEARCH TRENDS

BIOCHEMICAL STUDIES ON SOME BIOMARKERS OF XENOBIOTIC EXPOSURE

BIOCHEMISTRY RESEARCH TRENDS

Additional books and e-books in this series can be found on Nova's website under the Series tab.

BIOCHEMISTRY RESEARCH TRENDS

BIOCHEMICAL STUDIES ON SOME BIOMARKERS OF XENOBIOTIC EXPOSURE

VIVEK KUMAR GUPTA
AND
BECHAN SHARMA

nova
science publishers
New York

Library of Congress Cataloging-in-Publication Data

ISBN: 978-1-53619-381-7

Published by Nova Science Publishers, Inc. † New York

CONTENTS

List of Figures

List of Tables

PREFACE

Pesticides are known to cause the oxidative stress in mammalian system. These chemicals are widely used in the agricultural and health practices. The residual amounts of carbamate and organophosphate pesticides have been detected in the soil, water, vegetables, grains and other foods products. It has been reported that the exposure to the pesticide may induce toxicity in the exposed subjects which may manifest in the form of numerous adverse effects on health of the non-target living systems. The rise in pesticide applications has raised several environmental concerns. According to an estimate, only 2-5% of the total pesticides sprayed reach to the target and the rest reaches to different biotic and abiotic components of the environment through the pesticide drift.

Currently, the organophosphate and carbamate are being indiscriminately used to control the pests. These classes of pesticide have been shown to exert its toxic effects in various animal systems including the nervous system. Both of these groups of pesticides have a similar mode of action, affecting the nervous system by disrupting acetylcholinesterase activity. They disrupt the sodium/potassium balance of nerve fiber, following the forced them continuous conduction of nerve impulse which leads in developing several disease symptoms including weakness or paralysis. xiowever there is a marked difference between the modes of inhibition of AchE activity by cartap and malathion. The toxicity of organophosphate is mediated via covalent interaction of its phosphate group with the -OH group of serine residue present at the active site of acetylcholinesterase whereas carbamate interact at the same site non-covalently. Therefore, the organophosphate are irreversible inhibitors while carbamates are reversible inhibitors of acetylcholinesterase. In spite of these the above-mentioned pesticides have also been reported to induce the oxidative stress, and cause perturbations in energy metabolism, of the living systems.

The organophosphate particularly, the malathion is a pesticide commonly used to eradicate ectoparasites, rodents, and insects to conserve stored grain and to eliminate disease-inducing arthropods. On the negative side, it is one of the pesticides that exerts its diverse toxic effects through the inactivation of serine esterases, which leads to the over stimulation of cholinergic pathways. The organophosphate can reach to all the tissues, leading to several pathological aberrations due to their lipophilic nature and their simple and rapid intestinal assimilation, thereby causing pancreatitis, liver diseases, kidney injury, decrease in fertility and reproduction capability in both the humans and animals. Exposure to organophosphate have shown to influence xenobiotic biotransformation in liver through the generation of oxidative stress via production of excess amount of reactive oxygen species.

Carbamate has been reported to exert their toxic effects through the modulation of melatonin receptors and inhibition of acetylcholinesterase in the key organs of mammalian systems. Lipophilic nature of carbamate may render them with the ability to interact with the serum and tissue lipids. Carbamates have been reported to adversely influence the xenobiotic metabolism in hepatocytes. Cartap (being the member of thiocarbamate group of pesticides) is recognized as an analogue of nereis toxin. It is a neurotoxic substance isolated from the marine annelid *Lumbriconereis heteropoda*. However, not much information is available about its impact on the redox status into the organs and tissues of the mammalian systems. I addition carbamatesalso been reported to induce oxidative stress in different animal systems.

The chemicals enhancing the effect of other ones are called as synergists. The pesticides in combinations have been shown to induce synergistic effect by competing for a common target site. The exposure of mixture of pesticides is common in the aquatic environment, including lakes, river, streams, and other surface water that support aquatic life. Assessing the cumulative toxicity of pesticides in combination has therefore been a challenge for environmental health assessment. These two classes of chemicals inhibit activity of acetylcholinesterase and interfere with the cholinergic transmission. Since anticholinesterase agents have the common mode of action, the cumulative toxicity of the mixture can be estimated from the sum of the individual toxic potencies. The assumption of dose addition or concentration addition for mixtures of pesticides has also been demonstrated previously. But the synergistic effect of cartap and malathion has not been attempted so far. Further, the issues connected with acute poisoning, identification of long-term effects, susceptibility, and mechanisms of interaction are the areas of interest for the research in order to assess the risk quantitatively.

The exposure of living systems to different pesticides has been shown to induce oxidative stress. Oxidative stress reflects an imbalance between the systemic manifestation of reactive oxygen or reactive nitrogen species and the biological system's ability to detoxify the reactive intermediates or to repair the resulting damage. Disturbances in the normal redox state of cells can cause toxic effects through the production of peroxides and free radicals that damage all key components of the living cell, including proteins, lipids, and nucleic acids. The protective systems like oxidized glutathione/reduced glutathione, glutathione reductase, catalase, superoxide dismutase and peroxidase activities, ascorbate and α-tocopherol, are all recognized as potential antioxidants in the biological systems. The plant-based chemicals are also getting attention to be used as effective antioxidants.

Different xenobiotics can cause toxicity by acting on to different specific targets and lead to alterations in key biochemical parameters, known as biomarkers. However, the information regarding the specificity of biomarkers induced due to exposure of particular xenobiotics is still lacking. Biomarkers are the molecules that can be identified and monitored in different parts of the body. It refers to a broad subcategory of clinical signs. Biomarkers serve as indicators of a particular sensitivity of individuals to the effect of a xenobiotic or to the effects of a group of such compounds. Any variation from the normal level of a particular marker is indicative of abnormal function of the specific organ concerned, which probably may lead to the onset of a disease. The biomarkers serve as clinical and diagnostic tools and as surrogate end points in clinical research.

The increasing attention has been given towards the biomarkers of pesticides toxicity in mammalian systems. These markers are used to detect the adverse clinical impact of pesticides. The term biomarker includes almost any measurement showing an interaction between the environmental agent and the biological system, which could be biological, chemical, and physical. Biomarkers for monitoring pesticide exposures are typically divided into three broad categories: biomarkers of exposure, biomarkers of effect and biomarkers of susceptibility. These biomarkers provide information on the potential dose, absorbed dose and biologically effective dose of pesticide, an early stage in the development of disease, an indicator of an acquired ability to respond to the challenge of pesticide. These biomarkers are the measurable biochemical, physiological and the behavioral indices of animals such as inhibition of cholinesterase activity, β-glucuronidase, cytogenetic biomarkers (aberrations in chromosomal, micronuclei, sister chromatid exchange, single cell gel electrophoresis or comet assay), delta-aminolaevulinic acid dehydratase, hematological parameters and non-enzymatic biochemical parameters such as MDA and GSH. Some enzymatic antioxidative parameters include superoxide dismutase, catalase, glutathione-S-transferase, glutathione peroxidase, and glutathione reductase. The phosphatases, transferases, and cytochrome P450 have also been considered as effective markers of pesticide stress.

In order to alleviate the toxic impacts of various xenobiotics including the pesticides, the molecules derived from plants have shown great promises. The herbal medicines due to their low cost and least side effects have emerged as the richest source of antioxidants which have been exploited as a viable supplement with the numerous therapeutics against various diseases including neurotoxicity, hepatotoxicity, and cardiotoxicity. Most of the herbal products and/or herbal medicines including *A. vera* are the rich source of antioxidants. *A. vera* has been reported to contain large amount of various secondary metabolites and several natural antioxidant agents. There is no evidence available on the effects of whole herbs extracts of *A. vera* plant or an isolated phytochemical from it in alleviation of pesticide toxicity. *A. vera* is commonly used in folk medicine as an antidiabetic, anti-oxidative, anti-inflammatory, and antimicrobial agent. In fact, other important biological activities such as hypolipidemic, antiatherosclerosis, and anticarcinogenic activities of *A. vera* have also been reported.

The literature survey indicated that the pesticides belonging to carbamate and organophosphate groups have the potential to induce oxidative stress. However, the role of aqueous extract of *A. vera* has not been properly studied to demonstrate its ameliorative potential against the toxicity induced by these pesticides in single and combination in different organs of the mammalian systems. After extensive literature survey on the subject, an endeavor has been made to define the following objectives in order to fill up the existing lacuna of information about cartap and malathion induced oxidative stress in rat blood, brain and liver and its impact on neurotransmission as well as energy metabolism. Liver and blood were also being included in this study for establishing a more meaningful agreement. The present study was therefore undertaken to delineate the differential impacts of cartap and malathion on the biochemical and histological parameters in Wistar rats. In addition, the prophylactic effect of aqueous extract of *A. vera* leaves has also been evaluated.

The objectives of the presen study includes; to evaluate the xenobiotics induced oxidative stress, determination of activity of acetylcholinesterase and other esterases in the brain of rats exposed to xenobiotics, as potential biomarkers of neurotoxicity, evaluation of different enzymes of energy metabolism in rats exposed to xenobiotics, measurement of specific parameters concerning to the functions of kidney and liver in rats exposed to xenobiotics, evaluation of AchE isozyme profile to specifically understand the target of a xenobiotics in the brain of rats, Analysis of patterns of major biomolecules (proteins/DNA) in rats exposed to the xenobiotics for different treatment durations, Assessment of ameliorative impact of phytochemicals, if any, on xenobiotics.

This book is divided into eight sections. Section 1 contains the general introduction and background of the study. Section 2 contains extensive literature survey on the subject so as to present updated information. Section 3 demonstrate the main objectives of the study. Section 4 describes the various experimental design and procedures protocols and tools and techniques. Section 5 illustrated the data obtained from this study. Section 6 enumerates the discussion of the data and associated mechanisms. Section 7 includes the summary and conclusion of the entire study. Section 8 includes citations and references used in this work. This book contains many specific diagrams, illustrations and presentations in simple tables and figures which are self explanatory so as to make any reader well versed to this subject. In addition, the specific flow charts are added to simplify the presentation.

We hope that this publication would be of great use to graduates, postgraduates, researchers and faculty members who are studying in the field of plant based pcinciples for alleviating the pesticide toxicity and associated diseases. The tools and techniques explained in this title could be of immense use of all those working in this area. The authors would welcome the comments aand suggestions to improve the quality of the contents in the next edition.

Vivek Kumar Gupta, Ph.D.
Senior Research Fellow-ICMR
Department of Biochemistry
University of Allahabad
Prayagraj-211002,
U.P., India

&

B. Sharma, Ph.D.
Professor of Biochemistry (Cadre)
Department of Biochemistry
University of Allahabad
Prayagraj-211002,
U.P., India

ABOUT THE AUTHORS

Vivek Kumar Gupta has been awarded PhD in enzymology and molecular toxicology from Department of Biochemistry, University of Allahabad, Prayagraj. He was awarded the most prestigious Senior Research Fellowship of Indian Council of Medical Research-New Delhi in 2018. He has received a Brand ambassador-2018-19 award of Bentham Science Publishing. Mr. Gupta has also awarded a prestigious Young Scientist Award-2020. He has been awarded an Outstanding contribution in reviewing for Toxicology Reports (Elsvier)-2017. He has also received an International Travel Award by European Society for Virology to present his research work in the 4th Innovative Approaches for Antiviral Agents

Summer School, held at the University of Cagliari, Italy during September 24-28, 2018. He has awarded with travel grant to attend 3rd ACSTM and 6th ACSE Annual Conference in Deira Dubai, UAE during February 12-14, 2019. He has participated in several conferences and symposia and got exposure to scientific deliberations. He is a member of different scientific societies in India and abroad and has published 22 research papers in several peer reviewed journals of international repute. Mr. Gupta is on the panel of reviewers for 10 international journals and Editorial Board Member for one international journal.

Bechan Sharma is presently working as a professor/Ex-chairman, Department of Biochemistry, University of Allahabad, India, he completed all his higher education from Banaras Hindu University-Varanasi, India. He carried out his doctoral research at Central Drug Research Institute-Lucknow, India. Prior to joining at University of Allahabad, he worked at Sher Kashmir University of Agriculture Science and Technology-Srinagar, Kashmir (J&K) as an Assistant professor (Veterinary Biochemistry) and Dr. RML Avadh University Ayodhya-UP as a Reader (Biochemistry). The area of his research interests includes Molecular Biology, Tropical Diseases, Phytochemistry and Toxicology. With 32 years of teaching and 37 years of research experience in Biochemistry/Biotechnology/Environmental Sciences, he has served in different capacities for many academic and administrative bodies of many universities and institutions. He has received several awards honors and medals. He has produced 13 PhDs and mentored 7 postdoctoral research fellows. He is member/life meber of various national/international scientific societies and attended many symposia/conferences in indiaand abroad. He has published more than 225 research papers, 5 books, 46 book chapters, 5 editorials, 3 molecular methods and several popular scientific articles in many peer reviewed journals of national and international repute. He has one US patent on HIV-1 genome structure-based drug design to his credit. He is a member/life member of several national and international scientific societies. He is fellow of Academy of Environmental Biology (FAEB)-India and Bioved Research Society-Prayagraj He is a lead Guest editor of CMB (France), Natural Science (Biological control), and Natural Science (Research on enzyme); Chief editor of International Research Journal of Pharmacy and Pharmacology (AJPR, Academic journals), Journal of Biochemistry Research (JBR), Sky Journal of Biochemistry Research (SJBR), and Advances in Enzyme Research (AER). He is also working as a Guest editor/Associate editor/Executive editor/Editor member for 150 international journals and as honorary reviewer for over 175 scientific journals of international repute. He has worked as a visiting scientist in USA for 3 years and also at Italy, France, Thailand, Germany, Iran, Kuwait, Turkey and Brazil for 2-4 months where he has completed various research projects on different aspects of molecular biology of HIV-1, antiHIV-1 drug resistance and *in silico* drug design and development, gene silencing and toxicology. Prof. Sharma has been awarded ICMR's senior scientist fellowship 2014-2015 in biomedical sciences under Indo-USA joint research programme to work at NIEHS, NC, USA with Prof. Samuel Wilson on DNA repair. Recently, Prof Sharma ha sbeen included as a Life Member of World Virology Society, Sweden and in Indo-Japanese collaborative research initiativeon exploring plant-based molecules towards eradication of HIV/AIDS in India.

ACKNOWLEDGMENTS

The authors remember the almighty who gave the courage and strength, power and enthusiasm for completing the task assigned. We express our thanks and deep sense of gratitude to our well wishers, friends and family members for providing continued affection, encouragement, and cooperation which inspired us to fulfill the commitment. The completion of this project would not have been possible without their incredible support. The authors also extend their sincere thanks for the great academic opportunities and the basic amenities such as the library, subscription, and Internet facilities provided by the University of Allahabad. The financial assistance to one of the authors, Vivek Kumar Gupta, by University Grants Commission, New Delhi and Indian Council of Medical Research, New Delhi in the form of Research Fellowships is highly acknowledged.

We warmly acknowledge the hard work of the artists, typesetters, and other individuals including the contribution of the publication house, and technical persons, who made this book to appear in the final form. We are thankful to the honorable reviewers, who contributed their ideas, suggestions and critical comments on the manuscript.

Vivek Kumar Gupta, Ph.D.
Senior Research Fellow-ICMR
Department of Biochemistry
University of Allahabad
Prayagraj-211002, U.P., India

&

B. Sharma, Ph.D.
Professor of Biochemistry (Cadre)
Department of Biochemistry
University of Allahabad
Prayagraj-211002, U.P., India

Chapter 1

INTRODUCTION

Xenobiotics are commonly known as the foreign chemicals found within an organism that are not produced by them. The name xenobiotic stems up from the Greek words (Xenos) = foreigner, stranger and (Bios) = life, plus the Greek suffix for adjectives "tic". It includes mostly the synthetic chemicals or man-made chemicals (anthropogenic chemicals) as well as the molecules produced by other organisms reaching into a living system which do not produce it. Xenobiotics' introduction into any biological systems has been shown to generate free radical production which culminates into production of oxidative stress (Agrawal and Sharma, 2010; Gupta et al., 2016; Gupta et al., 2019).

Pesticides may be defined as chemicals that are able to control insects and pests including weeds. These chemicals are widely used in the agricultural and health practices. The residual amounts of carbamate (CAs) and organophosphate (OPs) pesticides have been detected in the soil, water, vegetables, grains and other foods products. It has been reported that the exposure to the pesticide may induce toxicity in the exposed subjects which may manifest in the form of numerous adverse effects on health of the non-target living systems. The rise in pesticide applications has raised several environmental concerns. According to an estimate, only 2-5% of the total pesticides sprayed reach to the target and the rest reaches to different biotic (animals and plants) and abiotic (water, air and soil) components of the environment through the pesticide drift (Agrawal and Sharma, 2010; Gupta and Sharma, 2015; Gupta and Sharma, 2016; Gupta et al., 2019).

Currently, the OPs and CAs are being indiscriminately used to control the pests. These classes of pesticide have been shown to exert its toxic effects in various animal systems including the nervous system (Jurewicz and Hanke, 2008). Both of these groups of pesticides have a similar mode of action, affecting the nervous system by disrupting acetylcholinesterase (AChE) activity (Colovic et al., 2013). They disrupt the sodium/potassium balance of nerve fiber, following the forced them continuous conduction of nerve impulse which leads in developing several disease symptoms including weakness or paralysis. The toxicity of OPs is mediated via covalent interaction of its phosphate group with the -OH group of serine residue present at the active site of AChE whereas CAs interact non-covalently. Therefore, the OPs are irreversible inhibitors while CAs are reversible inhibitors of AChE. In spite of these the above-mentioned pesticides has also been reported to induce the oxidative stress, and perturbations in energy metabolism, of the mammalian systems (Gupta and Sharma, 2015; Gupta and Sharma, 2016).

OPs particularly, the malathion [O,O-dimethyl-S-(1,2-dicarcethoxyethyl) phosphorodithioate] is a pesticide commonly used to eradicate ectoparasites, rodents, and insects to conserve stored grain and to eliminate disease-inducing arthropods (Babu et al., 2006). On the negative side, it is one of the pesticides that exerts its diverse toxic effects through the inactivation of serine esterases (Taylor et al., 1995), which leads to the over stimulation of cholinergic pathways (Bartling et al., 2007). The OPs can reach to all the tissues, leading to several pathological aberrations due to their lipophilic nature and their simple and rapid intestinal assimilation (Handy et al., 2002), thereby causing pancreatitis (Gokalp et al., 2005), liver diseases (Franco, et al., 2009), kidney injury (Afshar et al., 2008), decrease in fertility and reproduction capability (Selmi et al., 2018) in both the humans (Rothlein et al., 2006) and animals (Da Silva et al., 2006). Exposure to OPs have shown to influence xenobiotic biotransformation in liver (Lasram et al., 2009) through the generation of oxidative stress via production of excess amount of reactive oxygen species (ROS) (Yang et al., 2000).

On the other side, CAs has been reported to exert their toxic effects through the modulation of melatonin receptors (Popovska-Gorevski et al., 2017) and inhibition of acetylcholinesterase (Fukuto, 1990; Colovic et al., 2013) in the key organs of mammalian systems (Gupta et al., 1994). Lipophilic nature of CAs may render them with the ability to interact with the serum and tissue lipids. CAs have been reported to cause alterations in the levels of total serum lipids, glucose and protein levels in rats (Sadek et al., 1989). Some reports suggested that CAs adversely influenced the xenobiotic metabolism in hepatocytes (Pineiro-Carrero and Pineiro, 2004). Cartap being the member of thiocarbamate (TC) group of pesticides is the subclass of CAs with the basic chemical structure of cartap to be S,S-[2-(dimethylamino)-1,3-propanediyl] dicarbamothioate. Cartap is recognized as an analogue of nereis toxin (a neurotoxic substance isolated from the marine annelid *Lumbriconereis heteropoda*) (Ray, 1991). However, not much information is available about its impact on the redox status into the organs and tissues of the mammalian systems.

The chemicals enhancing the effect of other ones are called as synergists. The pesticides in combinations have been shown to induce synergistic effect (Cedergreen, 2014) by competing for a common target site (Cedergreen et al., 2017). The exposure of mixture of pesticides is common in the aquatic environment, including lakes, river, streams, and other surface water that support aquatic life (Gilliom, 2007). Assessing the cumulative toxicity of pesticides in combination has therefore been a challenge for environmental health assessment (Monosson, 2005). These two classes of chemicals (OPs and CAs) inhibit activity of AChE and interfere with the cholinergic transmission. Since anticholinesterase agents have the common mode of action, the cumulative toxicity of the mixture can be estimated from the sum of the individual toxic potencies. The assumption of dose addition or concentration addition for mixtures of anticholinesterase pesticides has also been demonstrated previously (Cedergreen, 2014). But the synergistic effect of cartap and malathion has not been attempted so far.

Oxidative stress reflects an imbalance between the systemic manifestation of reactive oxygen or reactive nitrogen species (ROS/RNS) and the biological system's ability to detoxify the reactive intermediates or to repair the resulting damage. Disturbances in the normal redox state of cells can cause toxic effects through the production of peroxides and free radicals that damage all components of the cell, including proteins, lipids, and DNA. Oxidative stress was shown to be induced by anthropogenic contaminants as persistent organic pollutants, heavy metals and also by toxins produced during massive blooms of cyanobacteria. The protective systems like oxidized glutathione/reduced glutathione, glutathione reductase, catalase, superoxide dismutase and peroxidase activities, ascorbate and α-tocopherol, are all recognized as potential antioxidants. The plant-based chemicals are also getting attention to be used as effective antioxidants (Agrawal and Sharma, 2012).

Different xenobiotics can cause toxicity by acting on to different specific targets and lead to alterations in key biochemical parameters, known as biomarkers. However, the information regarding the specificity of biomarkers induced due to exposure to particular xenobiotics is still lacking. Biomarkers are the molecules that can be identified and monitored in different parts of the body. It refers to a broad subcategory of clinical signs. Biomarkers serve as indicators of a particular sensitivity of individuals to the effect of a xenobiotic or to the effects of a group of such compounds. Any variation from the normal level of a particular marker is indicative of abnormal function of the specific organ concerned which probably may lead to the onset of a disease. The biomarkers serve as clinical and diagnostic tools and as surrogate end points in clinical research.

The increasing attention has been given towards the biomarkers of pesticides toxicity in mammalian systems. These markers are used to detect the adverse clinical impact of pesticides. The term biomarker includes almost any measurement showing an interaction between the environmental agent and the biological system, which could be biological, chemical, and physical (WHO, 1993). Biomarkers for monitoring pesticide exposures are typically divided into three broad categories: biomarkers of exposure, biomarkers of effect and biomarkers of susceptibility (WHO, 1993). These biomarkers provide information on the potential dose, absorbed dose and biologically effective dose of pesticide, an early stage in the development of disease, an indicator of an acquired ability to respond to the challenge of pesticide (Benford et al., 2000). These biomarkers are the measurable biochemical, physiological and the behavioral indices of animals such as inhibition of cholinesterase activity, inhibition of neuropathy target esterase activity, β-glucuronidase, cytogenetic biomarkers (aberrations in chromosomal, micronuclei, sister chromatid exchange, single cell gel electrophoresis or comet assay), delta-aminolaevulinic acid dehydratase, hematological parameters and non-enzymatic biochemical parameters such as MDA and GSH. Some enzymatic antioxidative parameters include superoxide dismutase, catalase, glutathione-S-transferase, glutathione peroxidase, and glutathione reductase. The phosphatases, transferases, and cytochrome P450 have also been considered as effective markers of pesticide stress (WHO, 1993).

To alleviate the toxic impacts of various xenobiotics including the pesticides, the molecules derived from plants have shown great promises. The herbal medicines due to their low cost and least side effects have emerged as the richest source of antioxidants which have been exploited as a viable supplement with the numerous therapeutics against various diseases including neurotoxicity (Gupta et al., 2015; Gupta and Sharma, 2017), hepatotoxicity (Jaiswal et al., 2017), and cardiotoxicity (Jaiswal et al., 2017). Most of the herbal products and/or herbal medicines including *A. vera* are the rich sources of antioxidants (Sanchez-Machado et al., 2017). *A. vera* with wide geographical distribution has been reported to contain large amount of various secondary metabolites and several natural antioxidant agents (Wang et al., 2008; Lopez et al., 2017). There is no evidence available on the effects of whole herbs extracts of *A. vera* plant or an isolated phytochemical from it in alleviation of pesticide toxicity. *A. vera* is commonly used in folk medicine as an antidiabetic, anti-oxidative, anti-inflammatory, and antimicrobial agent (Ghosh et al., 2011; Marco et al., 2015). In fact, other important biological activities such as hypolipidemic, antiatherosclerosis, and anticarcinogenic activities of *A. vera* have also been reported (Ghosh et al., 2011).

The biological monitoring is an important element of studies designed to assess the risk from pesticide exposure. The use of biomarkers as an integrated measure of xenobiotic exposure is increasing in assessment of risk. Thus, the issues connected with acute poisoning, identification of long-term effects, susceptibility, and mechanisms of interaction are the areas of interest for the research in order to assess the risk quantitatively. The extensive survey of literature indicated that the pesticides belonging to CAs and OPs groups have the potential to induce oxidative stress (Agrawal and Sharma, 2010). However, the role of aqueous extract of *A. vera* has not been properly studied to demonstrate its ameliorative potential against the toxicity induced by pesticides in single and combination in different organs of the mammalian systems. The present study was therefore undertaken to delineate the differential impacts of cartap and malathion on the biochemical and histological parameters in Wistar rats. In addition, the prophylactic effect of aqueous extract of *A. vera* leaves has also been evaluated.

Chapter 2

REVIEW OF LITERATURE

2.1. XENOBIOTICS

Xenobiotics are the substances that are non-self to the organism. Some of the xenobiotics are resistant to degradation either by chemical or biological processes. Xenobiotic substances are becoming an increasingly large problem, as they are relatively new substances and are very difficult to get metabolised in the living systems as they are hydrophobic innature and usually get associated with the fat bodies and hence deposited in different organs. Some examples of xenobiotics include synthetic organochlorides such as plastics and pesticides, or naturally occurring organic chemicals such as polyaromatic hydrocarbons, some fractions of crude oil and coal. However, it is reported that some microorganisms are capable of degrading many of the different complex xenobiotics found on the earth (Thornalley, 1990). The term xenobiotic is also used to refer the organs transplanted from one individual to another. For example, some researchers hope that heart and other organs could be transplanted from pigs to humans. Kidneys are currently the most commonly transplanted organ. Xenobiotic organs would need to be developed in such a way that they would not be rejected by the immune system.

Studies on how people transform the substances that they ingest began in the mid-nineteenth century, with chemists discovering that organic chemicals such as benzaldehyde, which could be oxidized and conjugated to amino acids in the human body. During the remainder of the nineteenth century, several other basic detoxification reactions were discovered, such as methylation, acetylation, and sulfonation. In the early twentieth century, work moved on to the investigation of the enzymes and pathways that were responsible for the production of these metabolites (Murphy, 2001). This modern biochemical research evolved from the identification of glutathione-S-transferases (Booth et al., 1961), followed by the discovery of cytochrome P450s (Omura and Sato, 1962), and the realization of their central role in xenobiotic metabolism (Estabrook et al., 1963; Estabrook, 2003). After the xenobiotics enter into the body systems through either of the entry routes such as contact to skin, respiration, mouth, injection or accidental ways, it undergoes series of physiological events known as absorption, distribution, metabolism, excretion and secretion.

2.2. Mobilization of Xenobiotics in the Body

The mobilization of xenobiotics takes place through a series of physiological processes such as absorption, distribution, metabolism, excretion and secretion.

2.2.1. Absorption

Absorption is the process by which toxicants cross body membranes and enter the blood stream or tissues (e.g., dermis). Factors affecting absorption include: (1) amount of toxicant present, (2) concentration of toxicant, (3) physical and chemical nature of toxicant such as (a) water partition co-efficient (high number indicates higher lipid solubility), (b) solid versus solution, (c) pKa value (highly acidic and highly alkaline compounds are less efficiently absorbed).

2.2.2. Mechanisms of Absorption

Structure of the cell membrane contains bilayer phospholipids, in which at one end have polar (charged and water-soluble) the phosphate group and another end a non-polar (uncharged and lipid soluble) lipid tails. The lipid bilayer contains lots of other types of molecules (e.g., proteins and carbohydrates) embedded in the membrane. Embedded molecules have various functions (e.g., receptors for external molecules, transport of materials across the membrane, connecting cells to each other or to other structures in the body like bone or basement membrane, recognition of self/non-self and enzymatic activity).

The rate of transport of a molecule across a membrane will correlate with the lipid solubility and size of the compound. The more a compound is lipid soluble, the more easily it will be able to penetrate the lipid bilayer of the cell membrane. In general, lipophilic compounds are more readily absorbed across membranes than hydrophilic ones. However, as molecules are absorbed into the body, they are typically carried away by the blood stream. If a toxicant is an organic acid or base, it tends to be absorbed by simple diffusion through the gastrointestinal (GI) tract. As water flows through pores in a membrane, any water-soluble molecules that are in solution and are small enough to fit through the pores will be transported along with the water. The absorption could also be carrier-mediated. It is therefore saturable and does not require energy. It can not operate against the concentration gradient.

The cell membrane possesses specific pumps that can actively transport molecules across membranes against concentration gradient. Active transport is an energy-requiring process. This transport is a saturable process, because it is limited by the number of active transport molecules. Thus, it can be overwhelmed by an excess of compound on one side of the membrane.

2.2.3. Routes of Absorption

2.2.3.1. Gastrointestinal (GI) Tract

The pH varies in the different sections of the GI tract, will affect the rate of absorption of chemicals. Numerous carrier mediated absorption processes occur in the GI tract (e.g., facilitated diffusion of glucose across the intestinal epithelium). Some chemicals are absorbed through the GI tract, and then transported to the liver, where they are excreted into the bile. These chemicals can then be reabsorbed and undergo multiple rounds of absorption and excretion. This can be highly significant if the molecules are hepatotoxic, because the liver is continually exposed to this toxic action. Some chemicals can be metabolized by the cells in the GI tract, or extracted and metabolized by the liver, then excreted into bile with or without prior biotransformation. These chemicals are essentially eliminated before they ever reach any other areas of the body besides the portal circulation between the GI tract and the liver.

2.2.3.2. Respiratory Tract

Toxicants absorbed by the lungs are usually gases or aerosols. The cells that make up the alveoli (pockets in the deeper portions of the lungs) are known as type I pneumocytes. The membranes of the alveoli are very thin and the capillaries are in close contact with the pneumocytes, so that the distance for a chemical to diffuse is very short. Molecules absorbed by the lungs are rapidly removed by the blood flow. The most important characteristics that affect absorption of aerosols in the lung are: particle size that affects the depth within the lung to which the molecule can penetrate. The smaller particles penetrate more deeply.

2.2.3.3. Skin

The rate-determining the dermal absorption of chemicals is the epidermis of skin (stratum corneum i.e., the upper most layer of the skin epidermis). This outer layer of skin, consisting of densely packed keratinized cells that have lost their nuclei and are biologically inactive, functions as an effective barrier to penetration. Abrasion of the skin can affect the rate of absorption of compounds across the skin.

2.2.4. Distribution of Xenobiotics

The initial phase of distribution is dominated by blood flow, whereas the final distribution depends largely on the affinity of a chemical for various tissues (e.g., lipophilic molecules will tend to accumulate in tissues with high lipid content).

2.2.4.1. Redistribution of Toxicants

The final distribution of a compound may not be the same as the initial distribution. It is not uncommon for redistribution to take place. For example, dioxin is highly lipophilic and accumulates in fat. A person may be exposed to dioxin and have a relatively high body burden, but no significant toxic effects because the chemical is essentially in storage. However, if the person does something that changes the proportion of fat in his body (e.g., goes on a diet), then dioxin can be mobilized out of the fat and can cause toxicity. Other examples were given above in the discussion on binding proteins.

The initial phase of distribution is determined primarily by blood flow. Therefore, a well-perfused organ such as the liver, may attain a high initial concentration of the chemical. Affinity is a natural attraction that tends to cause atoms or molecules to combine, or associate with each other. Some molecules have higher affinity for each other than do others. For example, a receptor molecule on the surface of a cell will have a high affinity for its ligand, but very low affinity for other molecules. Therefore, when both the receptor and the ligand are present, there is a higher probability that the two molecules will be associated than that they will be separated. Another example of affinity is found in the fact that lipid-soluble compounds will have a greater tendency to enter and remain in organs where there is a high concentration of fat (e.g., liver and brain), than organs that have relatively low-fat content (e.g., heart).

2.2.5. Factors Affecting Distribution

Blood flow, metabolism and excretion of the compound may varies between tissues. The more thoroughly perfused a tissue is (i.e., the more blood flows throughout the tissue), the greater the distribution will be within the tissue. The covalent attachment of a foreign molecule is likely to result in toxicity. If a compound binds non-covalently to a tissue component, this is usually reversible. The body contains numerous binding proteins, the purpose of which is to transport materials through the blood to specific sites, or to retain materials within specific tissues. Albumin-produced in the liver, is the primary binding protein in the blood, and is responsible for transport of many endogenous materials through the blood (e.g., bilirubin, uric acid, vitamin C and thyroxine). It also efficiently binds to many xenobiotics (e.g., antibiotics, barbiturates and digitonin). Transferrin-produced in the liver, is responsible for binding of iron. Metallothioneins bind metals and sequester them within tissues. Molecules that are bound to albumin or other binding proteins are not immediately available for transport into tissues, and therefore are unable to exert their pharmacologic/toxic effects.

The binding of chemicals to plasma proteins is important because severe toxic reactions can occur if a toxicant is displaced from plasma proteins, increasing the free fraction of the toxicant in to plasma. This is the cause of many known drug interactions. For example, when a drug is in the circulation, it is 99% bound to albumin, and only 1% is free. Only that 1% is free to have its pharmacologic effect. If another drug is administered, that has a higher affinity for albumin, and displaces 1% of the first drug from binding to the protein, that doesn't sound like a lot, but effectively you have doubled the free amount of the first drug. This can have severe consequences. For example, if a sulfonamide (which binds strongly to albumin) is administered concurrently with an antidiabetic drug, the sulfonamide can displace the antidiabetic drug and precipitate a hypoglycemic coma.

Bones and teeth tend to store chemicals that are similar to calcium (e.g., fluoride, strontium, and lead). Excessive consumption of fluoride (that's why they say not to swallow your fluoride toothpaste) results in incorporation of fluoride into the enamel of the teeth, producing a characteristic chalky appearance and severely weakening the teeth. Note that these "compartments" are not true physical compartments within the body. They are just a way of modeling the way that chemicals are distributed, and the extent of distribution, within the body.

Conceptually, the total body water of an organism may be divided into three compartments: (1) plasma water (water present in the blood), (2) interstitial water (water present outside of cells and not within) and (3) intracellular water (water present within the cells). For plasma water, if a chemical/drug has a calculated volume of distribution (V_d) that is approximately the same as the total volume of blood within an organism (~5 L for an adult human of about 70 kg body weight), then one might say that the chemical appears to be located primarily in the blood, and to have poor distribution to the tissues. Of course, this is not strictly the case; it is more likely that the chemical/drug is atleast partially absorbed into tissues, but it may be excluded from RBC.

2.2.6. Barriers to Distribution

2.2.6.1. Blood-Brain Barrier

A protective mechanism to protect the very sensitive cells of the central nervous system from toxicants in the blood. Less well-developed in immature individuals (e.g., newborns and infants). Tight junctions between the cells that make up the capillary endothelium of the brain tend to exclude water-soluble molecules from passing into the brain. Capillary endothelial cells contain active transport molecules of any molecules that do get into the cells that tend to excrete back into the blood stream. Capillaries in the CNS are surrounded by glial cells that form a lipid barrier around the blood vessels. Protein concentration in the interstitial fluid in the CNS is low, so there is less osmotic "pull" across the membranes.

2.2.6.2. Blood-Testis Barrier

A protective mechanism in the testis that is similar to that in the brain is that there are tight junctions between the endothelial cells.

2.2.6.3. Placental Barrier

It consists of several cell layers (almost 6) between the maternal and fetal circulatory vessels in the placenta. Placenta is thought to be an absolute barrier to drugs and other chemicals until Thalidomide disaster in 1960. The xenobiotics (such as ethanol and lead) can cross this barrier easily.

2.2.7. Xenobiotic Metabolism/Biotransformation Reaction

Many xenobiotics undergo chemical transformation when introduced into the human body. The end result of biotransformation is either alteration of the parent molecule, or conjugation of the parent molecule with endogenous substances in the body. Enzymes involved in biotransformation are termed as the drug metabolizing enzymes. Although this is a misnomer because many of the substrates are not drugs, the term is still commonly used. These reactions often act to detoxify poisonous compounds; however, in some cases, the intermediates in xenobiotic metabolism can themselves be the cause of toxic effects.

Xenobiotic metabolism is divided into three phases. The phase I biotransformation reactions involve oxidation, reduction, hydroxylation, transamination and epoxidation etc. The phase I enzymes such as cytochrome P450 oxidases lead reactive groups into xenobiotics. These modified compounds are then conjugated to polar compounds in phase II reactions. These reactions are catalyzed by transferase enzymes such as glutathione-S-transferases. Finally, the conjugated xenobiotics are further processed in phase III, before being recognized by efflux transporters and pumped out of the cells.

The reactions of these pathways are of interest as part of drug metabolism and as a factor contributing to multidrug resistance. The actions of some drugs on enzymes involved in xenobiotic metabolism are a common reason for lethal drug interactions. The body removes xenobiotics by xenobiotic metabolism. This consists of the deactivation and the secretion of xenobiotics, and happens mostly in the liver. Secretion routes are urine, faeces, breath and sweat. Hepatic enzymes are responsible for the metabolism of xenobiotics by first activating them (oxidation, reduction, hydrolysis and/or hydration of the xenobiotic) and then conjugating the active secondary metabolite with glucuronic or glutathione or sulphuric acid followed by the excretion. An example of a group of enzymes involved in xenobiotic metabolism is hepatic microsomal cytochrome P450. The enzymes that metabolize xenobiotics are very important for the pharmaceutical industry, because they are responsible for the breakdown of drugs.

The detoxification of endogenous metabolites such as aldehydes and peroxides can not be attained by system. However, since these compounds are few, it is possible for enzymatic system to use specific molecular recognition mechanism to identify and remove them. Examples of specific detoxification systems are the various antioxidant systems and the glyoxalase system. The removal of materials from the body takes place through various routes. There are several routes of excretion which includes bile-through the liver, urine-through the kidneys, feces-through the intestines, expired air-through the lungs, sweat-through the skin and saliva-through the mouth.

2.2.8. Significance of Xenobiotic Biotransformation Reactions in Toxicology

Biotransformation is a main portion of the removal pathway for numerous xenobiotic compounds. Biotransformation may consequence into either decrease or an increase in toxicity. Biotransformation may also result into the creation of several reactive metabolites. For example, Acetaminophen is generally processed in liver by sulfation and glucuronidation reaction to form a non-toxic compound. As the quantity of acetaminophen in liver is comparatively low, the Phase II pathways may process the compound which results into complete detoxification. If the acetaminophen amount becomes high to overpower the capacity of Phase II pathways, an alternate metabolic pathway which comprises the Phase I enzymes, becomes active. The Phase I reaction product is a highly reactive quinoneimine, which may form adducts with cellular proteins. The binding of the reactive intermediate to cellular macromolecules destroys the activity of those molecules, and can lead to compromised cell function and, ultimately, cell death. Another good example is metabolism of carbon tetrachloride. Carbon tetrachloride is metabolized by the cytochrome P450 system in the liver by abstraction of one of the four chlorine atoms.

This results into the formation of highly reactive trichloromethane radical, which recruits a cascade of lipid peroxidation by removing a hydrogen atom from phospholipids. Damage to the cell membrane causes loss of osmotic integrity, cell swelling and death. The drug metabolizing enzymes activity is dependent on many factors such as species, age, sex, genetics, organ, general health status, diet, and the previous exposure to other compounds.

2.2.8.1. Clearance (Cl)

The ratio relating the rate of elimination of a chemical from an appropriate reference fluid (usually plasma) to its concentration in the same reference fluid is called clearance. It has the units of flow rate in volume cleared per unit time (e.g., mL/min). A clearance of 100 mL/min of a chemical means that 100 mL of blood/plasma is completely cleared of the compound in each minute that passes.

2.2.8.2. Half-Life ($t_{1/2}$)

The time required for the concentration of a chemical in the plasma to decrease by 50%. This is a constant for all but zero-order kinetics. For example, for a compound eliminated by first-order kinetics, if the concentration at time 0 is 4 mg/L, and the $t_{1/2}$ is 6 hours, then at the end of 6 hours, the plasma concentration will be (0.5×4 mg/mL =) 2 mg/L, and at the end of the next 6 hours, the concentration will be (0.5×2 mg/mL =) 1 mg/L, and so on.

2.2.8.3. Tolerance to Xenobiotics

Organisms can also evolve to tolerate xenobiotics. An example is the co-evolution of the production of tetrodotoxin in the roughskinned newt and the evolution of tetrodotoxin resistance in its predator, the common garter snake. In this predator-prey pair, an evolutionary arms race has produced high levels of toxin in the newt and correspondingly high levels of resistance in the snake (Jakoby and Ziegler, 1990). This evolutionary response is based on the snake evolving modified forms of the ion channels that the toxin acts upon, so becoming resistant to its effects (Jakoby and Ziegler, 1990; Mizuno et al., 2003).

2.3. PESTICIDES

Pesticides are substances that control the insect and pests. The term pesticide includes herbicide, insecticides, nematicide, molluscicide, piscicide, avicide, rodenticide, bactericide, insect repellent, animal repellent, antimicrobial and fungicide. In broad-spectrum, pesticide is a chemical or biological agent that incapacitates, deters, discourages, or kills the pests. Along with these profits, pesticides also have some potential toxicity to humans and other species. For example, they are used to kill mosquitoes that can transmit deadly diseases like West Nile virus, yellow fever and malaria. Since 2000 BC, pesticides have been utilized the humans to protect their crops. The Rig Veda, which is about 4000 years old, mentions the use of poisonous plants for pest control (Rao et al., 2007). By the 15th century, toxic chemicals (mercury, arsenic and lead) were being used to kill pests. In 17th century, nicotine sulfate extracted from tobacco leaves were used as an insecticide. In the 19th century, the introduction of two more natural pesticides pyrethrum, and rotenone, were derived from chrysanthemums and the roots of tropical vegetables, respectively (Miller, 2002). By 1975, Organochlorines such as DDT, were substituted by organophosphates and carbamates. Later then, pyrethrins have become the dominant insecticide. In 1960s, it was revealed that DDT prevented many birds from reproducing. The agricultural use of DDT is now banned under the Stockholm Convention on Persistent Organic Pollutants, but it is still used in some developing countries to prevent malaria and other diseases. The use of DDT has been banned in most countries for their persistence in the environment and toxicity. They also kill bees, wasps or ants that can cause allergic reactions. Each use of a pesticide conveys some allied risk. Proper use of pesticide may decrease these associated risks to a level accepted by pesticide regulatory agencies such as the United States Environmental Protection Agency (EPA).

2.3.1. Types of Pesticides

Pesticides are often categorized according to the kind of pest they (1) control (e.g., herbicides, insecticides, fungicides, rodenticides, and pediculicides), (2) chemical structure (e.g., organic, inorganic, synthetic, or biological), although the distinction can sometimes blur) and (3) physical state (e.g., gaseous fumigant) (Gilden et al., 2010). Biopesticides include microbial pesticides and biochemical pesticides. Plant-derived pesticides have been evolving rapidly. These include pyrethroids, rotenoids, and nicotinoids (Kamrin, 1997). Pyrethroid is an organic compound which is similar to the natural pyrethrin produced by the flower of Pyrethrums. It is commercial household insecticide. The pyrethroids interfere the functions of voltage gated sodium channels in the axonal membrane.

2.3.2. Regulation of Pesticide

In 1910, the first legislation providing federal authority for regulating pesticides was passed (Goldman 2007). Although the U.S. Environmental Protection Agency was established in 1970 and amendments to the pesticide law in 1972 (Goldman 2007), pesticides use got increased by 50-fold since 1950 (Miller, 2002). Seventy-five percent of all pesticides in the world are used in developed countries, but their application in developing countries is ever on increase (Miller, 2004).

In Europe, European Union legislation has been approved banning the use of highly toxic pesticides including those which are reported to be carcinogenic, mutagenic, toxic to reproduction, endocrine-disrupting, persistent, bioaccumulative and toxic and measures have been approved to improve the general safety from pesticides across all European Union member states. The pesticide regulations vary among countries. To deal with these inconsistencies, the United Nations Food and Agriculture Organization adopted an International Code of Conduct on the Distribution and Use of Pesticides in 1985 to build voluntary standards of pesticide regulation for different countries (Willson, 1996). The code was updated in 1998 and 2002 (Programmes: International Code of Conduct on the Distribution and Use of Pesticides. UN FAO, 2008).

In addition to the Environmental Protection Agency (EPA), the United States Department of Agriculture (USDA) and the United States Food and Drug Administration (FDA) prepared set standards for the level of pesticide residues that is allowed in crops (Toth, 1996). The EPA looks at what the potential human health and environmental effects might be associated with the use of the pesticides. In addition, the U.S. EPA uses the National Research Council's four-step process for human health risk assessment: (1) hazard identification, (2) dose-response assessment, (3) exposure assessment and (4) risk characterization (US EPA, 2007).

2.3.3. Pesticide and Environment

The use of pesticides raised a number of environmental concerns and reach to the non-target species, air, water and soil (Miller, 2004). In addition, use of pesticide reduces biodiversity, contributes to pollinator decline, destroys habitat, and threatens endangered species (Miller, 2004). In order to decrease the negative impacts, it is desirable that pesticides be degradable or at least quickly deactivated in the environment. Such loss of pesticide toxicity is due to both innate chemical properties and environmental processes of the compounds (Sims and Sommers, 1986). For example, the presence of halogens within a chemical structure often slows down degradation of the compound (Wolt et al., 1996).

2.3.4. Pesticide Formulation with Their Combination

Pesticides may act as stomach, contact or systemic toxicants or even fumigants, depending upon their mode of action. However, in practice, pesticide formulations almost always occur in mixtures with other ones. Using combinations of pesticide formulations for agricularal practices is very common in developing nations with the rationale of the workers being the mistaken promise of "more is better." The reasons for combining the pesticides are:

1) To increase the effectiveness of one of the chemicals, called as synergism.
2) To provide the better control than that obtained from one pesticide.
3) To control the diferent types of pests economically with the single application of two of more pesticide mixtures.

However, the combined xenobiotics insult of two or more componennts of a pesticide mixtureruns the risk of dose addition or interaction as they might act along multiple system despite belonging to the same class. There are several other problems with the usage of pesticide combination are as follows:

1) Effectiveness of one or both compounds may be reduced, if the component pesticide are incompatible and act antagonistically to one another.
2) Precipitation may occur and choke the nozzle of application equipment.
3) Various types of phytotoxicity may occur, resulting in reduction of crop yield.
4) Some of the pesticides activate one another.

5) Excessive residues may occur as a result of mixing multiple pesticide formulations.

2.3.5. Abnormality due to Pesticide Exposure in Humans

Pesticides cause acute and delayed health effects in exposed people. Pesticide exposure can cause many adverse effects, ranging from simple skin and eyes irritation to severe effects such as affecting the nervous system, mimicking hormones which can cause reproductive problems and cancer. There is extensive evidence which proves that the organophosphate insecticide exposure is associated with neurobehavioral alterations (Mink et al., 2011). It has also been shown that the pesticides exposure may result into the fetal death, neurological and birth defects (Sanborn et al., 2007). Due to the poor regulation and safety precautions, about 99% of pesticide related deaths occur in developing countries (Goldmann, 2004). According to the World Health Organization and the UN Environment Programme it was estimated that each year 3 million agricultural workers in the developing countries suffer from severe poisoning of pesticides and of them about 18,000 die (Miller, 2004). A report suggested that 25 million labors in developing countries may suffer from mild poisoning of pesticide per year (Jeyaratnam, 1990).

2.3.6. Carbamates

The carbamates are the derivative of carbamic acid. The esters of carbamate are also called as urethanes. While the carbamate which functioned as acetylcholinesterase inhibitor is commonly referred to as carbamate insecticides. Due to their generally high selectivity for insect over mammalian acetylcholinesterase, the aldicarb and carbofuran (most potent compounds) are capable of inhibiting mammalian acetylcholinesterase at very low concentrations and pose a significant risk to the humans, especially when used in large amounts for agricultural practices. Other carbamate-based acetylcholinesterase inhibitors are also known for their higher toxicity to humans. Moreover, the carbamates are also used in human pharmacotherapy. The protease inhibitor (darunavir) used for HIV treatment also comprises a carbamate group. Carbamate insecticides also target melatonin receptors of humans, (Popovska-Gorevski et al., 2017) along with inhibiting the acetylcholinesterase (Colovic et al., 2013).

Thiocarbamates are a family of organosulfur compounds. They are the analogues of carbamates containing sulphur. There are two isomers of thiocarbamates: O-thiocarbamates and S-thiocarbamates.

2.3.6.1. Cartap

Cartap is a thiocarbamate insecticide. It is an analogue of (and a precursor for) the natural insecticide Nereis toxin, which works on the nervous system of the insects. Cartap is used to control the insects on crops. In 1967, cartap was first presented into the market in Japan. Its commercial names include Padan, Kritap, AG-Tap, Thiobel, and Vegetox. Its basic chemical structure is S, S-[2-(dimethylamino)-1, 3-propanediyl] dicarbamothioate. It is commonly used as a hydrochloride ($C_7H_{15}N_3O_2S_3HCl$). It has been categorized as a pesticide of high-effectiveness, low-toxicity, and low-residue used in rice and sugarcane fields. It is generally considered to be a safe compound with oral LD_{50} in the monkey of 100-200 mg/kg body weight (Liao et al., 2000). It was believed that it causes little skin or ocular irritation on exposure. However, subsequent studies have shown that ocular instillation can result in diaphragmatic contracture and death in rabbits (Liao et al., 2003).

Cartap has been recognized as an analog of nereis toxin which is a neurotoxic substance initially isolated from the annelid *Lumbriconereis heteropoda*. It is however believed that cartap is of a lower potency to nereis toxin. The other nereis toxin analogs that are in use are bensultap, thiocyclam and thiosultap. Its primary action was earlier considered to be by direct action on and inhibition of the postsynaptic nicotinic acetylcholine receptor (AChR) ion channels. A study published by Liao et al. (2000) challenges this belief based on studies of ocular cartap exposure in mice and its action on the phrenic nerve. The conclusion was that the effect on the diaphragm was not due to neuromuscular paralysis but instead due to the persistent diaphragmatic contraction resulting in respiratory failure.

Cartap may exert this effect primarily by promoting extracellular Ca^{2+} influx and induction of internal Ca^{2+} release. Cartap inhibits the [^{3}H]-ryanodine binding to the Ca^{2+} release channel in the sarcoplasmic reticulum in a dose-dependent manner. It was hypothesized that cartap-induced contracture was, to a minor extent, a result of inhibition of the sarcoplasmic reticulum Ca^{2+} pump protein Ca^{2+} ATPase. Inhibition of the Ca^{2+} ATPase would result in the unloading of calcium from the sarcoplasmic reticulum (Liao et al., 2000).

There have been a few reports of human toxicity in Japan and India as part of occupational exposure. Acute toxicity of the compound that has been described is nausea, vomiting tremulousness, salivation, spasms, dyspnea and mydriasis. Sodium dimercaptopropane sulfonate (DMPS) and sodium dimercaptosuccinate (DMS) were discovered to be effective antidotes for acute poisoning of insecticides like sodium ammonium dimethyl-2-(propane-1, 3-dithiosulfate) monohydrate, nereistoxin, and cartap in mouse models (Cao et al., 1990). These antidotes completely antagonized the respiratory depression caused by these compounds. At present the recommended antidotes for cartap poisoning are an intravenous injection of 100-200 mg of L-cysteine or an intramuscular injection of 20-60 mg of British Anti Lewisite (Dimercaprol; 2, 3-dimercapto propanol). It is possible that the effect of calcium binding of these antidotes is central to the antagonistic effects on respiratory depression in cartap toxicity.

2.3.6.2. Mechanism of Action of Cartap

Cartap is a nicotinic acetylcholine receptor (nAChR) channel blocker and has systemic insecticide action on stomach and contact. It causes paralysis by ganglionic blocking action on the CNS. Insects discontinue feeding upon contact, and die of starvation (Hartley and Kidd, 1987).

2.3.7. Organophosphates

Organophosphates (OPs) belong to a class of pesticides with the general structure $O{=}P(OR)_3$. They are the esters of phosphoric acid. The word organophosphate very often denotes a group of insecticides which act on the acetylcholinesterase. Today, organophosphates make up about 50% of the killing agents used among pesticides (Organophosphates: Background, Pathophysiology, Epidemiology, 2016*)*. Acetylcholinesterase break down the acetylcholine neurotransmitter, which sends out signals to other nerve endings in the body (Organophosphates: Background, Pathophysiology, Epidemiology, 2016*)*. OPs, inhibit this neuromuscular enzyme, which is broadly essential for normal function of the insects (Goodman, 2011). OPs affect this enzyme in varied ways, a principle one being through irreversible covalent inhibition (Peter et al., 2014). The human and animal toxicity of OPs make them a societal health and environmental concern (Goodman, 2011).

The US-EPA had banned for residential practices of OPs in 2001, but their agricultural use is still permitted (Goodman, 2011). For example, the most commonly used OP, such as malathion, (Bonner et al., 2007) has wide application in agriculture, residential land scaping and pest control programs. As of 2010, 40 such OPs were registered for use (Maugh and Thomas, 2010). The most commonly used OPs are malathion and parathion (Bonner et al., 2007), methyl parathion, diazinon, chlorpyrifos, phosmet, dichlorvos, fenitrothion, tetrachlorvinphos, azinphos-methyl, azamethiphos and Terbufos. OPs are rapidly degraded on exposure to sunlight, air, and soil, although their small amount can be detected in food and drinking water. When the pesticide degrades, it is broken down into several chemicals and hence contaminates the drinking water by moving through the soil to the ground water. It has been shown that prolonged exposure to OPs can lead to health problems, including increased risks for cardiovascular and respiratory disease, and cancer. In pregnant women, the exposure can result into premature births (Morello-Frosch et al., 2011).

The organophosphates are potent nerve agents, functioning by inhibiting the action of AChE. They are the most common reasons of poisoning, and are frequently used in agricultural areas. Organophosphosphates can be easily absorbed by ingestion, inhalation, and dermal contact. Their inhibitory effects on AChE lead to a pathological surplus of acetylcholine in body. Neurotransmitters such as acetylcholine are important in the development of a brain. Many organophosphates have been shown to exert neurotoxic effects on developing organisms, even at very low levels of exposure.

2.3.7.1. Malathion

Malathion is an organophosphate pesticide which functions as an inhibitor for the acetylcholinesterase. Malathion is widely used in agriculture, house hold practices and in public health to control pest. A malathion mixture with corn syrup was used in the 1980s in Australia and California to combat the Mediterranean fruit fly (Maugh and Thomas 2011). In Canada and the US, malathion was sprayed in many cities to combat west Nile virus. Today, Winnipeg is the only major city in Canada with an ongoing malathion adult-mosquito-control program (Morello-Frosch et al., 2011).

2.3.7.2. Mechanism of Action of Malathion

Malathion belonging to a diverse family of chemicals, has been reported to inhibit the activity of acetylcholinesterase. Upon uptake into the target organism, it binds irreversibly to several random serine residues on this enzyme. The resultant phosphoester group irreversibly inactivates the enzyme via covalent binding to the cholinesterases, which leads to quick acetylcholine accumulation at synapse.

2.4. Oxidative Stress and Reactive Oxygen Species

Reactive oxygen species (ROS) are chemical species containing reactive oxygen radical. Examples include peroxides, superoxide, hydroxyl radical, singlet oxygen, (Hayyan et al., 2016) and alpha-oxygen (Nachiappan et al.,

2010). The reduction of molecular oxygen (O_2) produces superoxide ($\bullet O^{-2}$) and is the precursor of most other reactive oxygen species (Turrens, 2003);

$$O^2 + e^- \rightarrow \bullet O^{-2}$$

Dismutation of superoxide produces hydrogen peroxide (H_2O_2) (Turrens, 2003):

$$2\,H^+ + \bullet O^{-2} + \bullet O^{-2} \rightarrow H_2O_2 + O_2$$

Hydrogen peroxide in turn may be partially reduced to hydroxyl radical ($\bullet OH$) or fully reduced to water (Turrens, 2003):

$$H_2O_2 + e^- \rightarrow HO^- + \bullet OH$$
$$2\,H^+ + 2\,e^- + H_2O_2 \rightarrow 2\,H_2O$$

In a biological context, ROS are formed as a natural byproduct of the normal metabolism of oxygen and have important roles in cell signaling and homeostasis (Devasagayam et al., 2004). However, during times of environmental stress, ROS levels can increase dramatically (Devasagayam et al., 2004).

2.4.1. Sources of ROS Production

2.4.1.1. Endogenous Sources

ROS are produced during a variety of biochemical reactions within the cell and within organelles such as mitochondria, peroxisomes, and endoplasmic reticulum (Han et al., 2001; Muller, 2000; Muller et al., 2007). Mitochondria convert energy for the cell into a usable form, adenosine triphosphate (ATP). The process of ATP production in the mitochondria, called oxidative phosphorylation, involves the transport of protons (hydrogen ions) across the inner mitochondrial membrane by means of the electron transport chain. In the electron transport chain, electrons are passed through a series of proteins via oxidation-reduction reactions, with each acceptor protein along the chain having a greater reduction potential than the previous. Normally, the oxygen is reduced to produce water; however, in about 0.1-2% of electron transfer chain, oxygen is instead prematurely and incompletely reduced to give the superoxide radical ($\bullet O^{-2}$), most well documented for Complex I and Complex III (Li et al., 2013).

Another source of ROS production is the electron transfer reactions catalyzed by the mitochondrial CYP450 systems in steroidogenic tissues (Hanukoglu et al., 1993). These CYP450 systems are dependent on the transfer of electrons from NADPH to CYP450. During this, the electrons leak out and react with molecular oxygen to produce superoxide radicals which can cause the cellular necrosis (Curtin et al., 2002; Albert et al., 2014).

2.4.1.2. Exogenous Sources

The formation of ROS can be stimulated by a variety of agents such as pollutants, heavy metals (Nachiappan and Muthukumar, 2010), tobacco, smoke, drugs, xenobiotics, or radiation. The radiation can produce harmful intermediates through the interacting water molecules, called radiolysis. Since about 60% of the human body contains water, the probability of radiation induced damage is quite high. In the process, water loses an electron and becomes highly reactive. Then through a three-step chain reaction, water is sequentially converted to hydroxyl radical, hydrogen peroxide, superoxide radical and ultimately oxygen. Another type of ROS is singlet oxygen, produced as a byproduct of photosynthesis in plants. In light reaction, chlorophyll may convert triplet to singlet oxygen (Laloi and Havaux, 2015). The increased production of singlet oxygen can result in cell death (Laloi and Havaux, 2015). Various substances such as carotenoids, tocopherols and plastoquinones contained in chloroplasts quench singlet oxygen and protect against its toxic effects. In addition to direct toxicity, singlet oxygen acts a signaling molecule (Laloi and Havaux, 2015).

2.4.2. Damaging Effects of Oxidative Stress

Effects of ROS generated due to the pesticide stress on cell metabolism are well documented in a variety of species (Nachiappan and Muthukumar, 2010). These include not only roles in apoptosis (programmed cell death) but also positive effects such as the induction of host defence (Conner et al., 2002; Rada and Leto, 2008) genes and mobilization of ion transport systems. This implicates them in control of cellular function. In particular, platelets involved in wound repair and blood homeostasis release ROS to recruit additional platelets to sites of injury. ROS are also implicated in mediation of apoptosis or programmed cell death and ischaemic injury. In general, harmful effects of reactive oxygen species on the cell are most often (Brooker, 2011): damage of nucleic acid, lipid peroxidation, oxidations of amino acids and the oxidative deactivation of specific enzymes by oxidizing the co-factors.

In aerobic organisms the energy needed for the normal biological functions is produced in the mitochondria through electron transport chain. In addition to the energy requirement, ROS are also produced which have the potential to cause cellular damage. Catalase and SOD may ameliorate the harmful effects of superoxide and hydrogen peroxide, by converting these compounds into oxygen and hydrogen peroxide. While ROS are produced as a product of normal cellular functioning, excessive amounts can cause deleterious effects (Patel et al., 1999).

2.4.2.1. Pathogen Response

In the mammalian host, ROS is induced as an antimicrobial defense. To highlight the importance of this defense, individuals with chronic granulomatous disease who have deficiencies in generating ROS, are highly susceptible to infection by a broad range of microbes including *Salmonella enterica*, *Staphylococcus aureus*, *Serratia marcescens*, and *Aspergillus* spp.

The exact manner in which ROS defends the host from invading microbe is not fully understood. More recently, a role for ROS in antiviral defense mechanisms has been demonstrated via Rig-like helicase-1 and mitochondrial antiviral signaling protein. Increased levels of ROS potentiate signaling through this mitochondria-associated antiviral receptor to activate interferon regulatory factor (IRF-3), IRF-7, and nuclear factor kappa-B (NF-κB), resulting in an antiviral state (West et al., 2011). Respiratory epithelial cells were recently demonstrated to induce mitrochondrial ROS in response to influenza infection. This induction of ROS led to the induction of type III interferon and the induction of an antiviral state, limiting viral replication (Kim et al., 2013). In host defense against mycobacteria, ROS play a role, although direct killing is likely not the key mechanism; rather, ROS likely affect ROS-dependent signalling controls, such as cytokine production, autophagy and granuloma formation (Deffert et al., 2014). Reactive oxygen species are also implicated in activation, anergy and apoptosis of T cells (Belikov et al., 2015).

2.4.2.2. Impairment of Cognitive Function

Memory capabilities decline with age, evident in human degenerative diseases such as Alzheimer's disease, which is accompanied by an accumulation of oxidative damage. Current studies demonstrate that the accumulation of ROS can decrease an organism's fitness because oxidative damage is a contributor to senescence. The results of this study have shown that the rats performed better after receiving the metabolites, suggesting that the metabolites reduced oxidative damage and improved mitochondrial function (Liu et al., 2002). Accumulating oxidative damage can then affect the efficiency of mitochondria and further increase the rate of ROS production (Stadtman, 1992). The accumulation of oxidative damage and their implications in aging depends on specific tissue where the injury is actually going on. The consequences advise that oxidative damage is accountable for age-related drop in functioning of brain. Older gerbils were shown to have greater levels of protein oxidation comparer to younger. This led researchers to conclude that oxidation of cellular proteins is potentially important for brain function (Carney et al., 1991).

2.4.2.3. Cause of Aging

According to the free radical theory of aging, oxidative damage initiated by reactive oxygen species is a major contributor to the functional decline that is characteristic of aging. While studies in invertebrate models indicate that animals genetically engineered to lack specific antioxidant enzymes (such as SOD), in general, show a shortened lifespan (as one would expect from the theory), the converse manipulation, increasing the levels of antioxidant enzymes, has yielded inconsistent effects on lifespan (though some studies in *Drosophila* do show that lifespan can be increased by the overexpression of MnSOD or glutathione biosynthesizing enzymes). Also contrary to this theory, deletion of mitochondrial SOD-2 can extend lifespan in *Caenorhabditis elegans* (Van Raamsdonk and Hekimi, 2009).

A study revealed an increased oxidative stress, reduced activities of antioxidant enzymes and DNA damage in the brain hippocampus and neocortex of prematurely aged rats than in control (Sinha et al., 2014). Numerous studies have shown that 8-OHdG (a product of ROS interaction with DNA) increases in different mammalian organs with age (Bernstein et al., 2008).

2.4.2.4. Male Infertility

Exposure of spermatozoa to oxidative stress is a major causative agent of male infertility (Aitken et al., 2012). Sperm DNA fragmentation, caused by oxidative stress, appears to be an important factor in the etiology of male infertility (Wright et al., 2014). A high level of the oxidative DNA damage 8-OHdG is associated with abnormal spermatozoa and male infertility (Guz et al., 2013).

2.4.2.5. Cancer

ROS are constantly generated and eliminated in the biological system and are required to drive regulatory pathways (Dickinson and Chang, 2011). Under normal physiological conditions, cells control ROS levels by balancing the generation of ROS with their elimination by scavenging system. But in oxidative stress, excessive ROS can damage cellular macromolecules, leading to the carcinogenesis.

Cancer cells exhibit greater ROS stress than normal cells do, partly due to oncogenic stimulation, increased metabolic activity and mitochondrial malfunction. ROS is a double-edged sword. On one hand, at low levels, ROS facilitates cancer cell survival since cell-cycle progression driven by growth factors and receptor tyrosine kinases (RTK) require ROS for activation (Irani et al., 1997) and chronic inflammation, a major mediator of cancer, is regulated by ROS. On the other hand, a high level of ROS can suppress tumor growth through the sustained activation of cell-cycle inhibitor (Ramsey and Sharpless, 2006; Takahashi et al., 2006) and induction of cell death as well as senescence by damaging macromolecules. In fact, most of the chemotherapeutic and radiotherapeutic agents kill cancer cells by augmenting ROS stress (Renschle, 2004; Toler et al., 2006). The ability of cancer cells to distinguish between ROS as a survival or apoptotic signal is controlled by the dosage, duration, type, and site of ROS production. Modest levels of ROS are required for cancer cells to survive, whereas excessive levels kill them.

Metabolic adaptation in tumours balances the cells' need for energy with equally important need for macromolecular building blocks and tighter control of redox balance. As a consequence, the NADPH production is significantly heightened, which functions as a cofactor to offer reducing power in numerous enzymatic processes and at the same time saving the cells from too much ROS formed during speedy proliferation. Cells counterbalance the detrimental effects of ROS by producing antioxidant molecules, such as reduced glutathione (GSH) and thioredoxin (TRX), which rely on the reducing power of NADPH to maintain their activities (Cairns et al., 2011).

The risk factors related with cancer, interacted the normal cells through ROS. ROS then activate various transcription factors such as nuclear factor kappa-light chain enhancer of activated B cells (NF-κB), activator protein-1 (AP-1), hypoxia-inducible factor-1α and signal transducer and activator of transcription 3 (STAT3), leading to expression of proteins that control inflammation; cellular transformation; tumor cell survival; tumor cell proliferation; and invasion, agiogenesis as well as metastasis. ROS also control the expression of various tumor suppressor genes such as p53, retinoblastoma gene (Rb), and phosphatase and tensin homolog (Gupta et al., 2012).

It has been assessed that ROS generated through the normal cellular metabolism can modify approximately 20,000 DNA bases per day in a cell. 8-oxoguanine is the most abundant among various oxidized nitrogeneous bases observed. During DNA replication, DNA polymerase mispairs 8-oxoguanine with adenine, leading to a G→T transversion mutation. The resulting genomic instability directly contributes to carcinogenesis. Cellular transformation leads to cancer and interaction of atypical PKC-ζ isoform with p47 phox controls ROS production and transformation from apoptotic cancer stem cells through blebbishield emergency program (Jinesh et al., 2016).

2.4.2.6. Cell Proliferation

Uncontrolled proliferation is a hallmark of cancer cells. The ROS have been revealed to increase cancer cell proliferation. The role of ROS in promoting tumor proliferation is further supported by the observation that agents with potential to inhibit ROS generation can also inhibit cancer cell proliferation (Gupta et al., 2012). Although ROS can promote tumor cell proliferation, a great increase in ROS has been associated with reduced cancer cell proliferation by induction of G2/M cell cycle arrest; increased phosphorylation of ataxia telangiectasia mutated (ATM), checkpoint kinase 1 (Chk 1), checkpoint kinase 2 (Chk 2) and reduced cell division cycle 25 homolog (CDC25) (Ames et al., 1983).

2.4.2.7. Cell Death

The death of a cancer cell can be in three means: necrosis, apoptosis and autophagy. Excessive ROS can induce apoptosis through both the extrinsic and intrinsic pathways (Ozben, 2007). In the extrinsic pathway of apoptosis, ROS are generated by Fas ligand as an upstream event for Fas activation via phosphorylation, which is necessary for subsequent recruitment of Fas-associated protein with death domain and caspase 8 as well as apoptosis induction (Gupta et al., 2012). In the intrinsic pathway, ROS function to facilitate cytochrome c release by activating pore-stabilizing proteins (Bcl-2 and Bcl-xL) as well as inhibiting pore-destabilizing proteins (Bcl-2-

associated X protein and Bcl-2 homologous antagonist/killer) (Martindale and Holbrook, 2002). The intrinsic pathway is also known as the caspase cascade and is induced through mitochondrial damage which triggers the release of cytochrome c. DNA damage, oxidative stress and loss of mitochondrial membrane potential lead to the release of the pro-apoptotic proteins mentioned above stimulating apoptosis (Maiuri et al., 2007). Mitochondrial damage is closely linked to apoptosis and since mitochondria are easily targeted there is potential for cancer therapy (Fulda et al., 2010).

The cytotoxic nature of ROS is a driving force behind apoptosis, but in even higher amounts, ROS can result in both apoptosis and necrosis, a form of uncontrolled cell death, in cancer cells (Hampton and Orrenius, 1997).

Numerous studies have shown the pathways and associations between ROS levels and apoptosis, but a newer line of study has connected ROS levels and autophagy (Gibson, 2010). ROS can also induce cell death through autophagy, which is a self-catabolic process involving sequestration of cytoplasmic contents (exhausted or damaged organelles and protein aggregates) for degradation in lysosomes (Shrivastava et al., 2011). Therefore, autophagy can also regulate the cell's health in times of oxidative stress. Autophagy can be induced by ROS levels through many different pathways in the cell in an attempt to dispose of harmful organelles and prevent damage, such as carcinogens, without inducing apoptosis (Scherz-Shouval and Elazar, 2007). Autophagic cell death can be prompted by the over expression of autophagy where the cell digests too much of itself in an attempt to minimize the damage and can no longer survive. When this type of cell death occurs, an increase or loss of control of autophagy regulating genes is commonly co-observed (Xie and Klionsky, 2007). Thus, once a more in-depth understanding of autophagic cell death is attained and its relation to ROS, this form of programmed cell death may serve as a future cancer therapy. Autophagy and apoptosis are two different cell death mechanisms brought on by high levels of ROS in the cells, however; autophagy and apoptosis rarely act through strictly independent pathways. There is a clear connection between ROS and autophagy and a correlation seen between excessive amounts of ROS leading to apoptosis (Scherz-Shouval and Elazar, 2007). The depolarization of the mitochondrial membrane is also characteristic of the initiation of autophagy. When mitochondria are damaged and begin to release ROS, autophagy is initiated to dispose of the damaging organelle. If a drug targets mitochondrion and creates ROS, autophagy may dispose of so many mitochondria and other damaged organelles that the cell is no longer viable. The high quantity of ROS leads to mitochondrial damage which may be a signal for apoptosis. The balance of autophagy within the cell and the crosstalk between autophagy and apoptosis mediated by ROS is crucial for a cell's survival. This crosstalk and connection between autophagy and apoptosis could be a mechanism targeted by cancer therapies or used in combination therapies for highly resistant cancers.

2.4.2.8. Tumor Cell Invasion, Angiogenesis and Metastasis

After growth factor stimulation of RTKs, ROS can trigger activation of signaling pathways involved in cell migration and invasion such as members of the mitogen activated protein kinase (MAPK) family-extracellular regulated kinase (ERK), c-jun NH-2 terminal kinase (JNK) and p38 MAPK. ROS can also promote migration by augmenting phosphorylation of the focal adhesion kinase (FAK) p130Cas and paxillin (Tochhawng et al., 2013). Both *in vitro* and *in vivo*, ROS have been shown to induce transcription factors and modulate signaling molecules involved in angiogenesis (MMP, and VEGF) and metastasis (Gupta et al., 2012).

2.4.2.9. Chronic Inflammation

Experimental and epidemiologic research over the past several years has indicated close associations among ROS, and chronic inflammation (Gupta et al., 2012). ROS induces chronic inflammation by the induction of COX-2, inflammatory cytokines (TNFα, interleukin 1 (IL-1), IL-6, chemokines (IL-8 and CXCR4) and pro-inflammatory transcription factors (NF-κB) (Gupta et al., 2012).

2.4.2.10. Cancer Therapy

Both ROS-elevating and ROS-eliminating approaches for cancer therapy have been established. Cancer cells with elevated ROS levels depend heavily on the antioxidant defense system. ROS-elevating drugs further increase cellular ROS stress level, either by direct ROS-generation or by agents that abrogate the inherent antioxidant system such as SOD inhibitor and GSH inhibitor. The result is an overall increase in endogenous ROS, which when above a cellular tolerability threshold, may induce cell death (Schumacker, 2006). On the other hand, normal cells appear to have, under lower basal stress and reserve, a higher capacity to cope with additional ROS-generating insults than cancer cells do (Trachootham et al., 2009). Therefore, the elevation of ROS in all cells can be used to achieve the selective killing of cancer cells. Radiotherapy also relies on ROS toxicity to eradicate tumor cells. Radiotherapy uses X-rays and γ-rays as well as heavy particle radiation such as protons and neutrons to induce ROS-mediated cell death and mitotic failure (Gupta et al., 2012).

2.4.2.11. Zymogram of AChE

The variable forms of the same enzyme are reffered as isozymes. They are highly homologous to each other but contain small differences in amino acid sequence. The term isozyme is not a precise one and can refer to different gene products, or different products of the same gene with alternative splicing. The term "isoenzyme/isozyme" should apply only to those multiple forms of enzymes arising from genetically determined differences in primary structure and not to those derived by modification of the same primary sequence. The term "multiple forms of the enzyme" is used as a broad term covering all proteins catalyzing the same reaction and occurring naturally in a single. In vertebrates a single gene encodes for AChE. However, the invertebrates possess variable number of AChE genes.

The AChEs are characterized by a very large heterogeneity firstly between different animal species (Vijayan and Oschowka, 1977) and secondly in different tissues of the same animal (Trevor et al. 1978). The heterogeneity was shown by the study of the physicochemical properties of the different isoenzymes, in particular using techniques of electrophoresis on starch gel (Erink et al. 1962) and polyacrylamide gel (Davis and Agranoff 1968; Vijayan and Brownson 1974; Michalek et al. 1978). AChE exists as mono-, di-, and tetramers of catalytic subunits (G1, G2, G4) with each unit containing active sites. These are designated as 'G' (globular forms) and 'A' (asymmetric collagen-tailed forms). All the molecular forms possess the same catalytic domain but distinct C-terminal peptides ($AChE_R$, $AChE_H$, $AChE_T$ and $AChE_S$).

$AChE_R$: Where R stands for 'read through'. The read through or the R transcripts or subunits of type R correspond to the hypothetical products of 'read through' transcripts, which retain the 'intronic' region that forms the last exon encoding the catalytic domain. $AChE_R$ molecular forms remain soluble and monomeric.

$AChE_H$: Where H stands for 'hydrophobic'. The hydrophobic or H transcripts or subunits of type H, are called so, because the C-terminal regions of these subunits are characterized by hydrophobic sequences. These hydrophobic regions consist of one or two cysteine residues, which establish disulphide bonds & result in a dimeric mature protein. The hydrophobic region majorly corresponds to a signal for a glycophosphotidyl (GPI) anchor.

$AChE_T$: Where T stands for 'tailed' forms $AChE_T$ is the only type of catalytic subunit that exists in all vertebrates producing major AChE forms in the brain and the muscle. $AChE_T$ generates multiple structures, ranging from monomers, dimmers to collagen-tailed and hydrophobic tailed forms. These tailed forms are catalytic tetramers associated with anchoring proteins, which help them attach to the basal lamina of the cell membrane.

$AChE_S$: Where S means 'soluble'. These $AChE_S$ subunits comprise of soluble monomeric forms of the enzyme and have been studied only in vertebrates.

In recent years, considerable interest has been directed toward the study of isoenzymes and their responses to physiological and pathological processes in the organism. Several investigators have reported the effects of pesticides on isoenzyme status of mammalian brain acetylcholinesterase (Christoff et al., 1966; Barron and Bernsohn, 1968; Eranko et al., 1962; Davis and Agranoff, 1968; Vijayan and Brownson, 1974). Use of high-resolution polyacrylamide gel electrophoresis in conjunction with quantitative procedures demonstrated that rat brain AChE may exist as three isoenzymes (Vijayan and Brownson, 1974). The alterations in AChE isoenzymes have been reported in several animal species including the rat (Bernsohn and Barron, 1964; Lagnado and Hardy, 1967; Iqbal and Talwar, 1971; Wilson et al., 1971). The present investigation was stimulated largely by the vast amount of information available on the anticholinesterase activities of carbamate and organophosphorous insecticides (Heath, 1961; Obrien, 1967). Most of the studies on mammalian toxicity of these compounds have been carried out using biochemical assays for AChE and very little effort has been expended toward the assessment of differential inhibition of multiple molecular forms of the enzyme. Such changes in other isoenzymes have been shown to occur in the absence of significant alteration in the total activity of the enzyme (Barth, 1970).

2.4.2.12. Genotoxicity and the Interaction of Pesticides with CYP Enzyme System

Evidences accumulating over the years suggested ceratin link between pesticides, chromosomal damage and the possibility of gene mutation (Kamiya et al., 2003). Among the variety of pesticides available, cartap (a carbamate) and malathion (an organophosphate) are frequently used in agricultural pracices. It has been shown to be a potential xenotoxic agent in mammalian systems. The pesticide induced oxidative stress (OS) involves inhibition of cytochrome P450 (CYP450), cytochrome reductase and cytochrome oxidase (Kamboj et al., 2007). These enzymes are involved in the oxidative metabolism of endogenous as well as exogenouschemicalsand produce their metabolite which are sometimes even more toxic than the parent molecule. Altered expression of CYP enzymes may be a consequence of pesticide induced xenotoxicity. In the view of important position of CYP enzyme system in the metabolism of xenobiotics and its role in pesticide toxicity, a detailed study of this effect is needed.

2.4.2.12.1. Cytochrome P450 (CYP450) Enzyme System

The cytochrome P450 (CYP enzymes) are a superfamily of heme containing enzymes, of which 2700 individual members are currently known to exist in nature (Lewis, 2004). They have been named after their characteristic absorption maximum (450nm), seen when the rediced form of the enzyme is bound to carbon monoxide (Omura and Sato 1962). The CYPs are able tomobilize a diverse group substrate, both endogenous and xenobiotics by catalyzing oxidative reactions. The CYP system is arguably our most important xenobiotic metabolizing enzyme system. Most of the CYP-catalyzed reactions lead to detoxification of xenobiotics, involving the formation the more polar metabolites that are more readily extractable. However, the CYPs can also activate prodrugs, or produce more toxic or reactive metabolites.

2.4.2.12.2. CYP Families

The families CYP1, CYP2 and CYP3 are primarily associated with the metabolism of exogenous compounds, whereas the other CYPs have mainly endogenous roles. The degree of similarity in the amino acid sequence divides the CYPs into families (>40% identical, e.g., CYP1 family), subfamilies (<55% identical, e.g., CYP1A subfamily), and individual enzymes with unique sequence e.g., CYP1A1 enzyme (Ingelman-Sundberg et al., 2000). In rats CYP1A2, CYP2B1, CYP2B2, CYP2C19, CYP3A1 and CYP43A are most abundantly expressed in liver and to a lesser extent in other organs such as brain, kidney and intestine (Hodgson et al., 1993).

2.4.2.12.3. Catalytic Mechanism of CYPs

CYPs most often catalyse the insertion of one of the atoms of molecular oxygen into the substrate being biotransformed, while the second atom of oxygen is reduced to water. The most frequently catalyzed reaction is hydroxylation, but the result of CYP-mediated catalysis can be e.g., a dealkylation, dehydrogenation, isomerization or carbon bond cleavage. The variety of CYP enzyme structures and the intrinsic reactivity of all their substrates explain the diversity of reactions catalyzed by CYPs (Werck-Reichhart and Feyereisen, 2000). The details of mechanism by which CYPs carry out all types of reactions are not fully understood. The best documented aspect is the hydroxylation reaction that is common to most CYPs. The selectivity of a substrate towards a particular CYP enzyme is based on the differences in the number and spatial disposition of the relevant complementary structure on both substrate and enzyme molecules. The lipophilic character and molecular mass of the substrate also play a role in substrate recognition. The *in vivo* biotransformation of a drug is often determined by the CYP with the highest affinity for the drug. Thus, the rate of elimination of drug can be largely determined by a single CYP enzyme or a combination of specific CYPs (Lewis, 2004).

2.4.2.12.4. Changes in the Activity of CYP

The level and activity of each CYP have been shown to vary from one individual to the next due to genetic, environmental and host factors. The environmental and host factors include medication (e.g., anticonvulsants, rifampicin and antifungal), foods (cruciferous vegetables and grapefruit), habbits (alcohol consumption and smoking), age, hormonal status and disease status (Sotaniemi et al., 2002).

2.4.2.12.5. Variation in the CYP Expression

CYP induction increases the capacity for metabolic detoxification. This is usually advantageous, protecting the individual against exposure to the xenobiotics, but can also be harmful, e.g., if the induced enzyme catalyzes the activation of a xenobiotic. The predominant mechanism in human is increased gene transcription, leading to an increase in the amount of induced enzyme.

The mechanism of CYP inhibition divides the inhibitors into reversible and irreversible (and quasi-irreversible) catagories. Reversible type of inhibition is dose dependent and and when the inhibitor is eliminated from the body, the normal metabolic function of the inhibited enzyme continues. Irreversible and quasi-irreversible inhibition requires metabolic activation of the inhibitor by the affected CYP enzyme before forming an enzyme inhibitor complex (Thummel and Wilkinson, 1998). Irreversible inhibitors are are commonly called catalysis dependent, mechanism based or suicide inhibitors. The covalent binding of the activated inhibitor to the enzyme lead to the irreversible inactivation of of the enzyme. The formed covalent bonds between an irreversible inhibitor and enzyme cannot be broken to regenerate the enzymatic activity. *in vivo* this sort of inhibition may start more slowly than reversible inhibition, but the final effect is usually more profound and is reversed only by the synthesis of new CYPs.

2.5. ANTIOXIDANT DEFENSE SYSTEM

The term antioxidant is generally used for the different groups of compounds (natural/synthetic) responsible for preventing the oxidative stress. The former, industrial antioxidants, have diverse uses: acting as preservatives in food and cosmetics, and being oxidation-inhibitors in fuels (Dabelstein et al., 2007). Antioxidants such as vitamin C terminate the chain reactions of oxidation. To balance the oxidative stress, plants and animals maintain complex systems of overlapping antioxidants produced internally or the dietary antioxidants vitamin C and vitamin E.

Antioxidants are classified into two broad divisions, depending on whether they are soluble in water (hydrophilic) or in lipids (lipophilic). In general, water-soluble antioxidants react with oxidants in the cell cytosol and the blood plasma, while lipid-soluble antioxidants protect cell membranes from lipid peroxidation (Sies, 1997). These complexes may be produced in the body or gained from diet (Vertuani et al., 2004). Some antioxidants are only found in a few organisms and these compounds can be important in pathogens and can be virulence factors (Miller and Britigan, 1997).

The comparative significance and relations between different antioxidants are a very complex, because various antioxidant molecules and antioxidant enzymes have the synergistic and interdependent effects (Chaudiere and Ferrari-Iliou 1999). The act of one antioxidant may consequently be governed by the proper functioning of other fellows of the antioxidant system (Vertuani et al., 2004). The amount of protection provided by any of the single antioxidant will also depend on its concentration, reactivity towards the particular reactive oxygen species and the status of the antioxidants (Vertuani et al., 2004).

2.5.1. Nonenzymatic Antioxidants (Thioredoxin and Glutathione Systems)

The nonenzymatic antioxidant system contains the 12-kDa protein thioredoxin and its companion thioredoxin reductase (Nordberg and Arner, 2001). The active site of thioredoxin contains two adjacent cysteines, as part of a highly conserved CXXC motif that can cycle between a reduced dithiol form and oxidized disulfide form. In its active state, thioredoxin acts as an efficient reducing agent, scavenging reactive oxygen species and maintaining other proteins in their reduced state (Arner and Holmgren, 2000). After being oxidized, the thioredoxin is restored by the action of enzyme thioredoxin reductase, using NADPH (Mustacich and Powis, 2000).

The glutathione system includes glutathione, glutathione reductase, glutathione peroxidases, and glutathione-S-transferases. This system is found in animals, plants and microorganisms (Meister and Anderson, 1988). Glutathione reductase helps in making oxidized glutathione into reusable reduced form. Glutathione peroxidase is an enzyme which contains four selenium-cofactors and catalyzes the breakdown of hydroperoxides and hydrogen peroxide. In addition, the glutathione-S-transferases show high activity with lipid peroxides (Sharma et al., 2004). These enzymes are at particularly high levels in the liver and also serve in detoxification metabolism (Hayes et al., 2005).

2.5.2. Enzymatic Antioxidants

As with the chemical antioxidants, cells are protected against oxidative stress by an interacting network of antioxidant enzymes (Sies, 1997). Here, the superoxide free radicals released by oxidative phosphorylation is first converted to hydrogen peroxide and then further reduced to water. This detoxification pathway is the result of multiple enzymes, with superoxide dismutases (SODs) catalysing the first step and then catalases (CAT) and various peroxidases removing hydrogen peroxide.

Superoxide dismutases (SODs) are a class of closely related enzymes that catalyze the breakdown of the superoxide anion into oxygen and hydrogen peroxide (Bannister et al., 1987). SODs are present in all cellular and in extracellular fluids (Johnson and Giulivi, 2005). These enzymes contain metal ion cofactors that, depending on the isozyme, can be copper, zinc, manganese or iron. In humans, the copper/zinc SOD is present in the cytosol, while manganese SOD is present in the mitochondrion (Bannister et al., 1987). There also exists a third form of SOD in extracellular fluids, which contains copper and zinc in its active sites (Nozik-Grayck et al., 2005). The mitochondrial isozyme seems to be the most biologically important of these three, since mice lacking this enzyme die soon after birth (Melov et al., 1998).

Catalase is a class of enzyme which uses either an iron or manganese as cofactor to catalyze the conversion of hydrogen peroxide to water and oxygen molecule (Chelikani et al., 2004). This protein is localized to peroxisomes in most eukaryotic cells (del Rio et al., 1992). Here, its cofactor is oxidised by one molecule of hydrogen peroxide and then regenerated by transferring the bound oxygen to a second molecule of substrate (Hiner et al., 2002). Despite its apparent importance in hydrogen peroxide removal, humans with genetic deficiency of catalase or mice genetically engineered to lack catalase completely, suffer few ill effects (Mueller et al., 1997).

Peroxiredoxins are peroxidases that catalyze the reduction of hydrogen peroxide, organic hydroperoxides, as well as peroxynitrite (Rhee et al., 2005). They are divided into three classes: atypical 2-cysteine peroxiredoxins, typical 2-cysteine peroxiredoxins and 1-cysteine peroxiredoxins (Wood et al., 2003). These enzymes possess same catalytic mechanism, in which a redox-active cysteine present in the active site is oxidized to sulfenic acid (Claiborne et al., 1999). Over-oxidation of this cysteine residue in peroxiredoxins inactivates these enzymes, but this can be reversed by the action of sulfiredoxin (Jonsson and Lowther, 2007). Peroxiredoxins appear to play a vital role in antioxidative metabolism, as the deficiency of this enzyme has shortened the lifespan of mice (Dietz et al., 2006).

2.6. Herbal Antioxidants

Herbs and spices are traditionally defined as any part of a plant that is used in the diet for their aromatic properties with no or low nutritional value (Davidson, 1999). However, more recently, herbs and spices have been identified as sources of various phytochemicals, many of which possess powerful antioxidant activity (Dragland et al., 2003). Thus, herbs and spices may have a role in antioxidant defense and redox signaling (Velioglu et al., 1998).

Interestingly, the antioxidants have been found to work together. Professor L. Packer of the University of California at Berkeley is one of the world's eminent antioxidant researchers. He and coworkers have demonstrated as how carotenoids interact with vitamins E and C. β-carotene was shown to protect low density lipoprotein against oxidative damage even when vitamin E levels are low (Yoshimi et al., 2011). Moreover, plant antioxidants such as phenols and bioflavonoids may potentiate vitamin antioxidants. For example, rutin, a bioflavonoid, potentiates vitamins C and E when taken together, producing more effective free radical scavenging activity (Yoshimi et al., 2011).

Antioxidants present in plants are the constituent nutrients with demonstrated free radical-scavenging abilities. Thus, in addition to α-tocopherol, carotenoids, ascorbate and zinc, plant-based medications may comprise polyphenols, flavonoids and flavoproteins. Further, some plants or their combinations in preparations may act as antioxidants by exerting free radical scavenging activity (Sawant et al., 2009). Further, they can be separated into many different groups on the basis of chemical structure, e.g., polyphenols, carotenoids thiols, retinoids, trace metals, terpenes, carbohydrates, tocopherols, degradation products of glucosinolates and others.

2.7. *Aloe barbadensis* Mill. (*A. vera*)

Herbal medicines derived from plants have been part of our traditional medicinal system since ancient human civilizations. Biological effects of *A. vera* have been extensively reviewed in the past few decades. The plant exhibited many pharmacological activities as antioxidant, antimicrobial, immune booster, antitumor, as well as hypoglycemic, and hypolipidemic agents. It helps in rapid wound healing. It has been used since ancient times as antiseptic, cosmetic and anti-inflammatory agent. Other traditional applications of *A. vera* concern treatment of burn injury, eczema, inflammation and fever, which continue to be studied. Various research studies are in pipeline to discover the potential of *A. vera* as immunity booster and treatment of AIDS. The new applications of *A. vera* are associated to food processing, such as coating of fruits, which is relatively safer and more effective compared to the post-harvest chemical treatments. Thus, *A. vera* preparations act as multipronged therapeutic agents against numerous health ailments. However, extensive research is needed to explore the therapeutic potential of *A. vera* in order to develop plant based novel drugs to treat many other health disorders. To ensure minimum or no drug contraindications about use of any of the *A. vera* products, the development of reliable testing protocols is required to assess the quality and quantity of the bioactive chemicals. This information might be fruitful for the better understanding of plant-based principles active against several diseases and to develop potential and safe therapeutics. *A. vera*-based principles could be further optimized and validated for their improved activities using bioinformatic tools. Since it is easy to grow this plant, its products can prove highly profitable, safe and more efficient against various diseases.

Among the genus *Aloe*, *A. barbadensis* Mill. (*A. vera*) and *A. arborescence* are the most common species. They generally grow in subtropical and tropical climates. The genus contains about 400 species (Eshun and He, 2004). This plant belongs to the Liliaceae family has been reported to be indigenous to South Africa (Reynolds and Dweck, 1999; Sahu et al., 2013; Manvitha and Bidya, 2014). Some researchers suggested that *A. vera* has a hybrid origin (Jones, 2000). According to the International Rules of Botanical Nomenclature, the scientific name for *A. vera* is *A. barbadensis* Mill. and *A. vera* (L.) Burm. F. is a synonym of the plant (Sahu et al., 2013). However, it is also known by several other names (Joseph and Raj, 2010, Ahlawat, and Khatkar, 2011). *A. vera* is a perennial, drought-resistant plant with thick, conical, green lance shaped, juicy, basal, sharp pointed and jagged and edged leaves. The fleshy leaves are covered with a thick layer of cuticle with whitish spots (Ahlawat and Khatkar; 2011). The plant can survive for more than 7 years without water (Joseph and Raj, 2010; Manvitha and Bidya, 2014). The leaves join in the stem forming structure like a rosette (Pal et al., 2013). Each plant normally has 12 to 16 leaves (25 and 30 cm long, and the 3-10 cm wide). The weight of a single mature leaf was reported to range from 1.5 to 2.0 kg (Ahlawat and Khatkar, 2011). The propagation is generally vegetative. The season of flowering and fruiting spans from October to April each year (Manvitha and Bidya, 2014). The long inflorescence of the plant has a large number of small bright yellow or red flowers (Manvitha and Bidya, 2014) and contains many seeds (Sahu et al., 2013). The plant has been used as an important ingredient in the traditional medicinal system since ancient time (over 2000 BC) for the treatment of various diseases (Rodriguez et al., 1988; Boudreau and Beland, 2006) prevalent in India, Japan, and China. Since biblical times, *A. vera* has been used for purgative purposes, healing of skin disorders and beauty treatments, and others (Hossain et al., 2013). In 1552 B.C., *A. vera* was mentioned as a laxative in the Egyptian Papyrus Ebers (Manvitha and Bidya, 2014). Besides this, Romans, Greeks, Arabs and Indians also recognized the healing properties of this plant (Joseph and Raj, 2010). In the 4th century B.C., Alexander The Great used *A. vera* for the treatment of his wounded soldiers. In 1920, *A. vera* was first cultivated for pharmaceutical delivery (Shelton 1991). Interestingly, among the species of *Aloe*, the *A. vera* has been considered as the most important species at commercial level (Eshun and He, 2004). In this scope, it is being cultivated in India, South Africa, Aruba, Bonaire, Venezuela, Haiti and the United States of America (Yeh et al., 2003). In the USA, *A. vera* has been grown mainly to supply the latex component of the leaf to the pharmaceutical industry (Lee et al., 2000). However, several species of *Aloe* have gained great popularity as therapeutic herb (Reynolds and Dweck, 1999; Radha and Laxmipriya, 2014). Many of the medicinal effects of *A. vera* leaf extracts have been attributed to the polysaccharides found in the inner leaf parenchymatous tissue (Ni et al., 2004). The clear pulp which is also known as gel is widely used in various medical, cosmetic and nutraceutical applications (Hamman, 2008).

The pharmacological actions of the gel were evaluated, and the results provided evidence that the plant contains anti-inflammatory, anti-arthritic, antibacterial and hypoglycemic activity (Lopez et al., 2017; Sanchez-Machad, et al., 2017), as well as anticancer, antioxidant, anti-inflammatory, immunomodulatory, hepatoprotective, antiulcer and antidiabetic properties (Reynolds and Dweck, 1999). The therapeutic potential of *A. vera* gel has been reported comprehensively (Habeeb et al., 2007) but the individual components should be explored. However, it is believed that these biological activities display the synergistic action of the compounds contained therein rather than a single chemical (Avijgan et al., 2014).

Many beneficial effects of this plant have been attributed to the polysaccharides present in the pulp (Rajasekaran et al., 2006). The phytochemicals present in it can be easily found in several cosmetics. Cosmetics, medicinal products and food supplements have been reported to be obtained from their mucilaginous tissues (derived from the center of succulent leaves of *A. vera*) and other parts of the plant (Salehi et al., 2018). Among naturally occurring anthraquinone derivatives, aloe-emodin and chrysophanol are the main compounds present in the gel (Tan et al., 2011; Mizuuchi et al., 2009; Lee et al., 2013).

Nowadays, the food, pharmaceutical and cosmetic industries are seeking new sources of natural compounds with interesting properties (Ribeiro-Santos, et al., 2015; Carvalho et al., 2015). Therefore, the optimization of the processes of plant harvesting, handling, transport, grinding, heating, dehydration, gel extraction and stabilization becomes pertinent to obtain more effective products. The optimization is also important to avoid changes in composition and properties (physiological and pharmaceutical) of *A. vera* products (Ramachandra and Rao, 2008). *A. vera* gel and whole leaf extract showed improved bioavailability of vitamins. This makes it possible to use *A. vera* as an excipient for of prolonged-release tablet formulations (Vinson et al., 2005; Jani et al., 2007). New products have also been developed from *A. vera*, such as the edible coating to preserve fruit (Valverde et al., 2005; Serrano et al., 2006; Martinez-Romero et al., 2013).

2.7.1. Phytochemistry (Nutrients and Bioactive Compounds) of the Plant

A. vera, a xerophytic plant, is adapted to live in sparse areas of water. The main feature of *A. vera* is its high-water content, stored in tissues, ranging from 99.00-99.50% (Hamman, 2008). The remaining solid material of 0.50-1.00% is reported as containing different potentially bioactive compounds/secondary metabolites including water- and fat-soluble vitamins, minerals, enzymes, monosaccharides, simple/complex polysaccharides (mucilages, gums), phenolic compounds (lignins, salicylates, tannins, quinolones), glycosides of phenolic (cyanogens, flavonoids, glucosinolates), terpenoids (carotenoids, iridoids, saponins, sesquiterpenes, steroids), peptides, resins and alkaloids (Femenia et al., 1999; Wang et al., 2008; Patel et al., 2012; Moriyama et al., 2016; Kar and Bera, 2018).

The leaves of *A. vera* constitute the rind (thick epidermis covered with cuticle surrounding the differentiated mesophyll in chlorenchyma cells and thinner walled cells) (Ramachandra and Rao, 2008). The rind was reported to be present at a fraction of 20-30% and the pulp was 70-80% compared to the weight of the whole leaf. In the dry powder of *A. vera*, the percentages of the lipid and protein contents in the rind were of 2.7% and 6.3%, respectively. The pulp of *A. vera* contains 4.2% and 7.3% of lipids and proteins, respectively (Femenia et al., 1999). The leaves secrete exudates called latex, which comes out from the pericyclic tubules. The leaves also contain hydroxyanthracene, anthraquinone, aloin A and B (Saccu et al., 2001). The polysaccharides of *A. vera* gel consists of the linear chains of glucose and mannose (Ni et al., 2004). The polymannose present in the gel of *A. vera* comprise linear chains of mannose molecules (Hutter et al., 1996). The acemannan has been reported as the main polysaccharide of *A. vera* gel.

The acemannan is constituted by polymers with repeating units of glucose and mannose in a ratio of 1:3 (Femenia et al., 1999; Chow et al., 2005) and located within the protoplast of the parenchymatous cells (Pugh et al., 2001; Chandegara et al., 2013). Aloin was reported to be present at levels of 0.66% as the main constituent of *A. vera* dry weight of leaf (Martinez-Romero et al., 2013). The hydrophilic aloin is converted into hydrophobic aloe-emodin in the intestine (Ernst, 2000). *A. vera* gel contains high water content (Rodriguez et al., 2010). A detailed list of the major phytochemicals is presented (Table 1 and 2) and summarized by several researchers (Lopez et al., 2017; Sanchez-Machad, et al., 2017). The phytochemistry and pharmacology of *A. vera* roots have not been studied sufficiently, which deserves more attention.

2.7.2. Biological Applications of *A. vera*

The biological activity of *A. vera* has been attributed to the presence of several active secondary metabolites (Eshun and He, 2004; Reynolds and Dweck, 1999; Boudreau and Beland, 2006). More than 200 bioactive chemicals have been found in *A. vera* gel (Lakshmi and Rajalakshmi, 2011). It is more likely that the biological activities of leaf extracts are due a synergistic action of the compounds, rather than a single compound (Hamman, 2008). Several therapeutic and biological indications are attributed to *A. vera*. The bioactive components of *A. vera* are being used in several treatments such as in skin diseases, radiation injuries, wound healing, gastrointestinal problems, dysentery, diarrhoea, burns, constipation, coughs, headaches, arthritis, immunodeficiency, ulcerative colitis, (Langmead et al., 2004), hepatoprotective (Gupta et al., 2019), antiseptic, antimicrobial, anti-inflammatory, antioxidant, hemostatic, antidiabetic, anti-obesity and anticancer agent (Radha and Laxmipriya, 2014; Kar and Bera, 2018). Recently, *A. vera* has been exploited for skin substitutes as nanofibrous structures. In addition, the incorporation of *A. vera* in this structure enhanced the fibroblast proliferation. This fabricated scaffold has been reported to have good biodegradability, drug delivery and antibacterial activity. *A. vera* based nanofibers have been reported to be used for skin tissue engineering applications (Baghersad et al., 2018).

2.8. Laboratory Rats

A laboratory rat is a rat of the species *Rattus norvegicus domestica* which is bred and kept for scientific research. While less commonly used for research than mice, rats have served as an important animal model for research in psychology and biomedical science (Vandenbergh 2000). A rat strain is a group of animals in which all members are as nearly as genetically identical. In rats, this is accomplished through inbreeding. Using such population of animals, it is possible to investigate the roles of genes. By contrast, outbred populations are used when identical genotypes are unnecessary or a population with genetic variation is required, and are usually referred to as stocks rather than strains (International Committee on Standardized Genetic Nomenclature for Mice/Rat Genome and Nomenclature Committee, 2016*)*). Scientists have bred many strains of rats specifically for experimentation. The common strains are:

2.8.1. Wistar Rats

The Wistar rat is an outbred albino rat, and was developed at Wistar Institute in 1906 for use in biomedical research. It was notably the first rat developed to serve as a model organism. Mostly, laboratory rats are descended from the wistar rats (Clause, 1998). The Wistar rats are categorized by its long ears, wide head, and the tail length which is always less than the length of body.

2.8.2. Long-Evans Rat

The Long-Evans rats are outbred developed by Long and Evans in 1915 by crossing several Wistar females with a wild gray male. These rats are white with a black or brown hood. They are utilized as a multipurpose model organism, frequently in behavioral and obesity research.

2.8.3. Sprague Dawley Rat

The Sprague Dawley rat is an outbred multipurpose breed of albino rat used extensively in medical and nutritional research (Hsu and Lai, 2007). Its main advantage is its calmness and ease of handling. This rat breed was first formed by the Sprague-Dawley farms in Madison, Wisconsin, in 1925. These rats normally have a longer tail in proportion to their body length than Wistar rats.

2.8.4. Biobreeding Rat

Biobreeding diabetes-prone rat is an inbred strain that spontaneously develops autoimmune type 1 diabetes. The Biobreeding rats are used as an animal model for Type 1 diabetes. The strain re-capitulates many of the features of human type 1 diabetes, and has contributed greatly to the research (Mordes et al., 2004).

2.8.5. Brattleboro Rat

The Brattleboro rat was established in West Brattleboro, Vermont. It has a naturally occurring genetic mutation that makes specimens unable to produce the hormone vasopressin, which helps control kidney function.

2.8.6. Hairless Rat

Hairless rats provide advantage to the researchers with valued information about the genetic kidney diseases and compromised immune systems. It is estimated that there are over twenty-five genes that cause recessive hairlessness in laboratory rats (Kim et al., 2004). The lack of this organ severely compromises their immune system, infections of the respiratory tract and eye increasing the most dramatically (Festing et al., 2004). Fuzzy rats were identified in 1976 in a Pennsylvanian lab. Shorn rats were bred from Sprague Dawley rats in Connecticut in 1998 (Moemeka et al., 1998).

2.8.7. Lewis Rat

The Lewis rats were established by M. Lewis from Wistar rats in the 1950s. Characteristics include albino coloring, docile behavior, and low fertility. The Lewis rat suffers from several spontaneous pathologies and their current research applications include transplantation research, induced arthritis and inflammation, experimental allergic encephalitis, and STZ-induced diabetes (Charles River Laboratories, 2013).

2.8.8. RCS Rat

The Royal College of Surgeons rats were the first animal with inherited retinal degeneration. This genetic flaw was identified in the year 2000 to be a mutation in the gene MERTK. This mutation results in defective retinal pigment epithelium phagocytosis of photoreceptor outer segments (D'Cruz et al., 2000).

2.8.9. Shaking Rat Kawasaki

The Shaking rat Kawasaki (SRK) is an autosomal recessive mutant rat that has a short deletion in the RELN (Reelin) gene (Kikkawa et al., 2003) resulting into the lowered Reelin protein expression, which is essential for proper lamination of cortex and the development of cerebellum. This rat is well-known stocks developed from Wistar rats.

2.8.10. Zucker Rat

The Zucker rats were raised to be a genetic animal model for investigation on hypertension and obesity. They are named after Lois M. Zucker and Theodore F. Zucker, pioneer researchers in the study of the genetics of obesity. There are two types of Zucker rat: a lean Zucker rat; and obese Zucker rat (Kurtz et al., 1989; Takaya et al., 1996; Kava et al., 1990).

2.8.11. Knockout Rats

A knockout rat is a genetically modified rat with single gene turned off by directed mutation. Knockout rats are the significant tools for learning the function of a gene which would be helpful in drug discovery and development. The production of knockout rats became technically feasible in 2008 by the members of the Knock Out Rat Consortium (KORC). Knockout rat model is being used for the study of hypertension, diabetes, Parkinson's and Alzheimer's disease.

After extensive literature survey on the subject, an endeavor has been made to define the following objectives for present dissertation in order to fill up the existing lacuna of information about cartap and malathion induced oxidative stress in rat blood brain and liver and its impact on neurotransmission as well as energy metabolism via acetylcholinesterase ad lactate dehydrogenase toxicity, respectively. An attempt was also made in this study to explore the possibility of prevention of cartap and malathion induced toxicity using natural antioxidants. Liver and blood were also being included in this study for establishing a more meaningful agreement.

2.9. OBJECTIVES

1) To evaluate the xenobiotics induced oxidative stress.
2) Determination of activity of acetylcholinesterase and other esterases in the brain of rats exposed to xenobiotics, as potential biomarkers of neurotoxicity.
3) Evaluation of different enzymes of energy metabolism in rats exposed to xenobiotics.
4) Measurement of specific parameters concerning to the functions of kidney and liver in rats exposed to xenobiotics.
5) Evaluation of AChE isozyme profile to specifically understand the target of a xenobiotics in the brain of rats.
6) Analysis of patterns of major biomolecules (proteins/DNA) in rats exposed to the xenobiotics for different treatment durations.
7) Assessment of ameliorative impact of phytochemicals, if any, on xenobiotics induced perturbations into rats.

Chapter 3

MATERIALS AND METHODS

3.1. CHEMICALS

Malathion was procured from Hindustan Insecticides Ltd. (a Government of India enterprise) New Delhi-India in liquid form having 50% w/w purity. Cartap hydrochloride was obtained from the local market in the semisolid form with 50% w/w purity of Dhanuka Agritech Limited, Gurgaon, Haryana, India, formulated in technical association with Sumitomo Chemical Company Ltd., Japan. Acetylthiocholine iodide (ATI) and hydrogen peroxide (H_2O_2) were purchased from TCI, Japan. Triton X-100, ethylenediaminetetraacetic acid (EDTA), pyrogallol, sodium acetate, sodium bicarbonate, sodium carbonate, sodium dihydrogen phosphate, sodium hydrogen phosphate and sodium hydroxide were obtained from Merk India Ltd. 1-chloro-2,4-dinitrobenzene (CDNB), Follin's reagent, glutathione, glycerol, NADH, sodium pyruvate, sodium chloride and tris-base, reduced glutathione (GSH) and bovine serum albumin (BSA) were purchased from Sisco Research Laboratories Pvt. Ltd., Mumbai, India. Alcohol was procured from Changshu Yanguan. Paraffin wax, potassium dichromate, sulphuric acid and xylene, were from Merk India Ltd. All other chemicals used were of analytical grade. The pre-stain molecular weight marker protein (Bio-Rad Laboratories), and 30% Acrylamide stock (30:1

acrylamide: bisacrylamide) were from Bio-Rad Laboratories. Bromophenol Blue was purchased from Thermo Fisher Scientific. Tris base was obtained from SRL. Glycine, glycerol, isopropanol and HCl from Merk. β-mercaptoethanol, TEMED, ammonium persulfate, SDS, coomassie brilliant blue R250, glacial acetic acid and methanol were procured from Sigma-Aldrich. The protein mini gel cassettes (Bio-Rad) were used for the electrophoresis.

3.2. Animals

The male Wistar rats, aged between 8 weeks, weighing 120±10 g was purchased from Central Drug Research Institute (CDRI)-Lucknow, India. All the animals were acclimatized at ambient temperature and environmental conditions in polypropylene cages in the standard laboratory conditions for 15 days. Each cage contains six rats in the laboratory under ambient environmental conditions. The animals were allowed to free access of water and food with 12 h light and dark. The experimental procedures were strictly designed and performed according to the guidelines of Institutional Animal Ethical Committee (IAEC), University of Allahabad, Prayagraj (CPCSEA Registration number: 839/GO/Re/04/CPCSEA, 1/10/2004).

3.3. Identification and Collection of *Aloe vera* Leaves

The aerial part i.e., leaves of the *A. vera* (L.) Burm.f. plant belonging to the family Asphodelaceae, sub-family- Asphodeloideae was selected for the studies. The young leaves of *A. vera* were obtained from Roxburgh Botanical Garden, University of Allahabad, Prayagraj. The fresh leaves of the *A. vera* plant freed from foreign organic and non-organic matter. The genus and species of the plants were identified and authenticated by an eminent Professor of Taxonomy, Department of Botany, University of Allahabad, India. The leaves were shade dried at room temperature.

3.4. Preparation and Storage of Plant Extract

Following which the shade dried *A. vera* leaves were powderized followed by the preparation of extract (1:10 w/v ratio of *A. vera* leaf powder and double distilled water). The obtained aqueous extract thus obtained was then concentrated at 30^0C, which was then freeze dried for complete removal of solvent and stored at room temperature. The aqueous extract thus obtained was then lyophilized for the complete removal of solvent and stored at -20^0C.

3.5. Calculation for Percent Water and Dry Matter Content

The water content is determined following the method of (Badreddine, 2016). 10 g of *A. vera* leaves were freshly weighed and dried in an oven at 50^0C for 24 h. The dry matter was removed from the oven. After cooling in a desiccator, the dried sample was weighed.

$$\text{Water content (\%)} = [(W_1 - W_2)/W_1] \times 100$$

W_1 stands for = weight of the sample before drying; whereas W_2 = stands for weight of the sample after drying.

Thus, dry matter content was extracted from water content as shown in the formula below:

$$\text{Dry matter content (\%)} = [100 - \text{water content (\%)}]$$

3.6. Screening of Phytochemical Components in the Extract

The phytochemical evaluation of *A. vera* extract was carried out in order to investigate the presence of total flavonoids, saponins, tannins, terpinoids, steroids, phenols, quinones, carbohydrates and alkaloids. The freshly prepared extract was subjected to the standard methods for analyses of different phytochemical constituents.

3.6.1. Test for Flavonoids

To 1.0 ml of aqueous extract of *A. vera*, 1.0 ml of 10% lead acetate aqueous solution was added and mixed. The formation of a yellow precipitate was taken as a positive test for flavonoids.

3.6.2. Test of Saponins

5.0 ml of aqueous extract of *A. vera* was mixed vigorously with 5.0 ml of distilled water in a test tube and warmed. The formation of stable foam, honey comb in shapes, was considered as an indication for presence of saponins.

3.6.3. Test for Tannins

About 2.0 ml of the aqueous extract of *A. vera* was stirred with 2.0 ml of distilled water. To it, few drops of $FeCl_3$ solution (5% w/v) were added. The formation of a green precipitate was an indication for the presence of tannins.

3.6.4. Test for Terpenoids

2.0 ml of the *A. vera* extract was dissolved in 2.0 ml of chloroform and evaporated to dryness. 2.0 ml of concentrated sulphuric acid was then added and heated for about 2.0 min. A grayish colour indicates the presence of terpenoids.

3.6.5. Tests for Steroids

I. A red colour produced in the lower chloroform layer confirms the presence of steroids when 2.0 ml of the aqueous extract of *A. vera* was mixed with 2.0 ml of chloroform followed by addition of 2.0 ml of concentrated sulphuric acid in a test tube.

II. The development of a greenish colour confirms the presence of steroids when 2.0 ml of the the aqueous extract of *A. vera* was mixed with 2.0 ml of chloroform and treated with 2.0 ml sulphuric and 2.0 ml acetic acids.

3.6.6. Tests for Anthraquinones (Borntrager's Test)

3.0 ml of aqueous extract of *A. vera* was mixed with 3.0 ml of benzene, filtered and 5.0 ml of 10% ammonia solution was added to the filtrate. The mixture was vortexed. The presence of a pink colour in the ammonical (lower) phase indicates the presence of free anthraquinones.

3.6.6.1. Tests for Carbohydrates (Molisch's Test)

3.0 ml of aqueous extract of *A. vera* was added to 2.0 ml of Molisch's reagent and the resulting mixture was shaken, then 2.0 ml of concentrated H_2SO_4 was poured carefully down the side of the test tube. A violet ring at the interphone indicates the presence of carbohydrate.

3.6.6.2. Test for Alkaloids

3.0 ml of aqueous extract of *A. vera* was mixed with 3.0 ml of 1% HCl on a steam bath. Mayer's and Wagner's reagents were then added to the mixture. Turbidity of the resulting precipitate was taken as an evidence for the presence of alkaloids.

3.7. Phytochemical Evaluation of Extract

The phytochemical evaluation of *A. vera* extracts were carried out in order to investigate the presence of total phenolics, flavonoids and flavonols contents as following:

A. Determination of total phenolic content: The total phenolic content in the modified folin-ciocalteau reagent method (Wolfe et al., 2003). An aliquot of the extract was mixed with 5ml FCR (previously diluted with water 1:10 v/v) and 4 ml (75 g/l) of sodium carbonate. The tubes were vortexed for 15 sec and allowed to stand for 30 min at 40^0C for color development. The optical absorbance was measured at 765 nm using spectrophotometer. Samples of extracts were evaluated at a final concentration of 0.1 mg/ml. Total phenolic content was expressed as mg/g gallic acid equivalent (mg/g).
B. Determination of total flavonoids: Total flavonoid content was determined using the method of (Ordon et al., 2006). A volume of 0.5 ml 2% $AlCl_3$ in ethanol solution was added to 0.5 ml of sample solution. After one hour of incubation at room temperature, the optical absorbance was measured at 420 nm. Extract samples were evaluated at final concentration of 0.1 mg/ml. Total flavonoid content was calculated as quercetin equivalent (mg/g).
C. Determination of total flavonols: Total flavonols content was estimated using the method of Kumaran and Karunakaran (2007). To 2.0 ml of sample standard, 2.0 ml of 2% $AlCl_3$ in ethanol and 3.0 ml (50g/l) sodium acetate solutions were added. The optical absorption at 440nm was read after 2.5h incubation at 20^0C. Extract

samples were evaluated at a final concentration of 0.1 mg/ml. Total flavonol content was calculated as quercetin equivalent (mg/g).

3.8. ANTI-OXIDANT ACTIVITY ASSAY USING 2, 2- DIPHENYL-1-PICRYLHYDRAZYL (DPPH)

The DPPH radical scavenging activity assay was carried out as described by Siatka and Kasparova (2010). The solution of DPPH (0.1 mM) in methanol was prepared and 1.0 ml of this solution was added to 1.0 ml of different concentrations (25, 50, 100, 250, 500 and 1000 μg/ml) of plant extract prepared in double distilled water. These solution mixtures were kept in dark for 30 min. After 30 min incubation, the absorbance was measured at 517 nm using UV-visible double beam spectrophotometer and ascorbic acid was served as positive control. The degree of DPPH decolourization from purple to yellow color indicated the scavenging effectiveness of the extract. A blank was prepared without adding extract. Ascorbic acid (1%, w/v) at various concentrations was used as a standard. The low value of the absorbance of the reaction mixture indicates the higher free radical scavenging activity. The capability to scavenge the free radical by DPPH was calculated using the following equation.

$$\text{DPPH scavenge effect (\%)} = [(Ac-As)/Ac] \times 100$$

Where, Ac and As are the absorbance values of the control and the sample, respectively.

3.9. EXPERIMENTAL DESIGN FOR ANIMAL CARE AND TREATMENT

The animal treatment with subacute doses of cartap, malathion and the mixture of cartap as well as the malathion and their amelioration by aqueous extract of *A. vera* leaves was attempted. The experimental animals (male

Wistar rats) were divided into eight groups. The experimental animals or male Wistar rats aged 8 weeks were divided into eight groups and each group containing six rats. The detailed experimental design is as follows:Group 1 was served as control (C*), had free access to rat feed and water only; Group 2 was plant extract (*A. vera*) treated control (PE), received 420 mg kg^{-1} body weight; Group 3 was cartap treated (C), received 10% LD_{50} (29 mg kg^{-1} body weight) of cartap; Group 4 was malathion treated (M), received 10% LD_{50} (29 mg kg^{-1} body weight) of malathion; Group 5 was treated with the mixture of cartap and malathion (C+M), received a mixture of 5% LD_{50} (14.5 mg kg^{-1} body weight) of each pesticide; Group 6 contained the animals treated with the plant extract (*A. vera*) i.e., 420 mg kg^{-1} body weight followed by cartap (10% $LD_{50,}$ i.e., 29 mg kg^{-1} body weight) treated (PE+C). Group 7 contained the treatment of plant extract (*A. vera*) (420 mg kg^{-1} body weight) followed by malathion (10% $LD_{50,}$ i.e., 29 mg kg^{-1} body weight) treated (PE+M). Group 8 received 420 mg kg^{-1} body weight of *A. vera* extract followed by the treatment of a mixture of both the pesticides (5% LD_{50};14.5 mg kg^{-1} body weight of each cartap and malathion).

The oral treatment was given for all groups up to 15 days at regular interval of 24 h. During this period, the clinical symptoms and body weight were also monitored. At the end of treatment, all animals were euthanized and sacrificed as per the guidelines of Institutional Animal Ethical Committee (IAEA), University of Allahabad, India. At the end of treatment, all animals were euthanized and sacrificed as per the guidelines of Institutional Ethical Committee, University of Allahabad, Prayagraj-India (CPCSEA Registration No.: 839/GO/Re/04/CPCSEA, 1/10/2004).

3.10. Assessment of Whole Blood on Hematological Parameters and Preparation of Blood Sample

After sacrificing the animals, the blood from the rat's heart was collected out in sterile heparinized blood vials. The blood was analyzed for complete blood count (CBC) using CBC machine. The blood was also processed for erythrocyte, and serum isolation. The blood samples were also collected in non-heparinized tubes and processed for the separation of serum. The serum was used for biochemical investigations. The packed red blood cells (PRBCs) were stored at 4^0C and the serum was either used for biochjemical analysis or stored at -20^0C for use when ever needed.

3.11. Preparation of Blood Serum

The blood was collected from each animal in a sterile blood collecting vial not coated with any anticoagulant. Blood was centrifuged at 1000xg at 4^0C and serum was stored for biochemical estimations at -20^0C. Packed RBCs were prepared by washing erythrocytes in isotonic saline.

3.12. Preparation of Ghost Cells/Packed Red Blood Cells (PRBCs) from Blood and Extraction of AChE from PRBCs Membrane

Packed red blood cells (PRBCs) were prepared by washing pelleted erythrocytes in isotonic saline buffer. PRBCs were used for the estimation of acetylcholinesterase (AChE). The PRBCs was into anticoagulant coated vial and was centrifuged at 3000xg for 5 min at 4^0C for the removal of plasma. The erythrocytes were washed thrice with cold phosphate buffer saline (20 mM phosphate buffer containing 274 mM NaCl and 5.4 mM KCl; pH 7.4) and centrifuged. After that, the PRBCs were resuspended in hypotonic phosphate buffer of 5 mM Tris-HCl containing 1 mM EDTA (pH 7.4) and kept at 4^0C for overnight incubation. Then the red blood cells (RBCs) were centrifuged for 8 times with the hypotonic and hypertonic phosphate buffer washing alternatively at 9000xg for 10min at 4^0C to get the ghost cell (erythrocyte membranes). This preparation (ghost cells after the loss of hemoglobin) was treated with Triton X-100 (0.2%, v/v; prepared in phosphate buffer 50 mM, pH 7.4) to solubilize membrane bound AChE enzyme. The ghost cells were homogenized by applying 3 strokes, each of 30 sec in the presence of precooled 50mM phosphate buffer (pH 7.4) containing 0.2% Triton X-100 (v/v) to produce homogenate at 4^0C. The homogenate was centrifuged at 10,000xg for 10 min at 4^0C and the supernatant was used for the AChE enzyme assays.

3.13. Animal Scarification and Sampling of Tissues

After sacrificing the animals, the tissue from all the key organs of the rats were dissected out, washed with PBS and stored at -80^0C. All the biochemicals and molecular estimations were done with these tissues.

3.14. Extraction of AChE from Rat Brain Tissue

The brain of rat was removed and washed with ice cold saline (0.9% NaCl) blotted to dry. For the estimation of acetylcholinesterase activity, the tissues were homogenized (10% w/v) in 50 mM phosphate buffer, pH 7.4 containing 0.2% Triton X-100 (v/v), at 4^0C using homogenizer. The remaining tissues were stored at -80^0C. The homogenates were kept in the capped tubes and centrifuged for 30 min at 4^0C with intermittent gentle shaking at 10000xg. The supernatant was collected in precooled Eppendorf tubes and the pellet was discarded. This preparation was used for both the assay of AChE activity as well as total protein content estimation.

3.15. Preparation of Tissue Homogenate in Sucrose

The homogenate (10%, w/v) was prepared in ice cold 0.25 M sucrose solution and centrifuged at 9000xg for 30 min at 4^0C. The supernatant was separated by gentle decantation of centrifuged homogenate of tissue and used for assay of antioxidant and other enzymes and estimation of certain biomolecules levels.

3.16. Non-enzymatic Antioxidant Assays

3.16.1. Estimation of Reduced Glutathione (GSH) Level

The total thiol content in the tissue homogenate and rat blood serum was determined by the method of Ellman et al., (1959) modified by Sedlak and Lindsay (1968). The optical density of the yellow color complex reaction mixture present in the supernatant was measured at 412 nm. In 2.0 ml microcentrifuge tube the reaction mixture contains 105 µl 0.2 M tris HCl buffer pH 8.2, 20 µl 0.01 M DTNB, 825 µl absolute methanol and 50 µl homogenate was mixed and incubated for 15 min with intermittent shaking and centrifuged at 800xg for 15 min. The result was expressed as µg GSH/mg protein. GSH standard was used for the calculation.

3.16.2. Estimation of Malondialdehyde (MDA) Level

Lipid peroxidation was measured in the blood serum of rats following the method of Niehaus & Samuelsson (1968) and the results were expressed as nmol MDA/mg protein using 1.56×10^{5} M^{-1} cm^{-1} as an extinction coefficient.

3.17. Enzymatic Antioxidant Assays

3.17.1. Assay of Superoxide Dismutase (SOD, EC 1.15.1.1) Activity

The activity of superoxide dismutase (SOD, E.C. 1.15.1.1) was measured by using the method of Marklund and Marklund (1974). It is spectrophotometric measurement of optical density of colored complex involving pyrogallol auto-oxidation at 412 nm for 3 min at the interval of 30 sec. One unit of the enzyme activity was expressed as 50% inhibition of auto-oxidation of pyrogallol min^{-1}.

3.17.2. Assay of Catalase (CAT, EC 1.11.1.6) Activity

The catalase (CAT, EC 1.11.1.6) activity was measured according to the method of Beers and Sizer (1952) by measuring the decrease in the absorbance for H_2O_2 consumption at 240 nm at the interval of 30 sec for 3 min. One unit of CAT activity was defined as µmoles of H_2O_2 decomposed min^{-1} using molar extinction coefficient of H_2O_2 (43.6 M^{-1} cm^{-1}).

3.18. Activity Assays for the Enzymes of Glutathione Redox System (GST and GPx)

3.18.1. Assay of Activity of Glutathione-S-Transferase (GST, EC 2.5.1.18)

The specific activity of glutathione-S-transferase (GST, EC 2.5.1.18) was estimated by following the method of Habig et al. (1974). The change in optical density of the reaction mixture was recorded at 340 nm at the interval of 30 sec for 3 min and the results were expressed as µmol/ml/min/mg protein. The activity of GST was calculated using the molar extinction coefficient ($9.6x10^3$ $µmol^{-1}$ cm^{-1}).

3.18.2. Assay of Glutathione Peroxidase (GPx, EC 1.11.1.9) Activity

The GPx activity in RBCs hemolysate was determined by the method of Pagalia and Valentine (1967). The rate of oxidized glutathione formation was measured spectrophotometrically at 340 nm with NADPH. The enzyme activity was expressed as IU/mg Hb.

3.19. Assay of Acetylcholinesterase (AChE) Activity

The AChE activity was assayed by following the method of Ellman et al. (1961). The reaction mixture (3.0 ml) in quartz cuvette having 1 cm path length contained 0.50 mM of ATI, 0.5 mM of DTNB, and 50 mM phosphate buffer (pH 7.4). The change in optical density was measured at 412 nm for 3 min at each interval of 30 sec. The AChE activity was calculated using extinction coefficient $13.6 \times 10^3\ M^{-1}\ cm^{-1}$ and expressed as μmoles of acetylthiocholine (ATI) hydrolyzed $ml^{-1}\ min^{-1}$ or units (U). The catalytic activity is measured by the increase of the yellow anion, 5-thio-2-nitrobenzoate, produced due to reaction of thiocholine with 5,5-dithio-bis-(2-nitrobenzoic acid) (DTNB). The specific activity of enzyme was expressed in IU mg^{-1} of protein. The enzyme assays were performed on UV-visible double beam spectrophotometer (Thermo Scientific Spectroscan UV 2700). The assay system without substrate or enzyme was considered as a substrate or enzyme blank, respectively, and any change in absorbance min^{-1} recorded in this condition was subtracted from the experimental observations.

3.20. Activity Assay for Lactate Dehydrogenase (LDH) as a Marker of Energy Metabolism System

The activity of LDH in cell free extract of brain, liver and serum was measured by the method of Horecker and Kornberg (1948). The 3.0 ml of reaction mixture contained 0.2 M Tris-HCl buffer, pH 7.4 (1.0 ml), 0.1 M KCl (0.15 ml), 50 mM sodium pyruvate (0.15 ml), 2.4 mM NADH (0.20 ml), suitably diluted enzyme and distilled water. The enzyme activity was monitored as decrease in the absorbance at 340 nm for 3 min. The reaction mixture without enzyme protein served as a control in this assay system. The specific activity of the enzyme was presented in IU/mg enzyme protein using extinction coefficient 6.22x10^3 μmol/L/min. The optical absorbance of UV radiation by the reaction mixture was monitored on UV-visible double beam spectrophotometer (Thermo Scientific Spectroscan UV 2700).

3.21. Activity Assay for Phosphatases as a Marker of Cell/Tissue Necrosis

The activity of acid phosphatase (ACP, EC 3.1.3.2) and alkaline phosphatase (ALP, EC 3.1.3.1) was measured according to the method of Goldberg and Barka (1962). The change in absorbance was recorded at 405 nm for 3 min at the interval of 30 sec. The specific activity of the enzyme was presented in μmol/ml/min/mg protein or IU/mg enzyme protein using. The reaction mixture without enzyme protein served as a control in this assay system. The optical density of the colour developed after reaction was monitored by UV-visible double beam spectrophotometer (Thermo Scientific Spectroscan UV 2700).

3.22. Activity Assay for Transaminases (SGOT and SGPT) as a Marker of Liver Function

The activities of aspartate transaminase (AST) (EC 2.6.1.1) known as serum glutamic oxaloacetic transaminase (SGOT), and alanine transaminase (ALT) (EC 2.6.1.2) formerly called as serum glutamate-pyruvate transaminase or serum glutamic-pyruvic transaminase (SGPT) in the serum as well as in the post nuclear supernatant (PNS) were monitored by the method of Reitman and Frankel (1957). The suitable amount of plasma enzyme protein or PNS of liver extract was incubated with 0.25 ml of substrate (aspartate and α-ketoglutarate for AST and alanine and α-ketoglutarate for ALT in phosphate buffer pH 7.0) for 30 min. 0.25 ml of DNPH was used to stop the reaction and kept for 20 min at room temperature. After incubation, 2.5 ml of 0.4 N NaOH was added and vortexed gently. The optical density was monitored at 510 nm by using the UV-visible double beam spectrophotometer (Thermo Scientific Spectroscan UV 2700). The reaction mixture without enzyme protein served as a control in each case.

3.23. Estimation of Uric Acid in Serum

The quantitative estimation of uric acid was determined using the Erba Mannheim kit (dynamic extended stability with lipid clearing agent modified Trinder method) in serum of rats. The optical density of the reaction mixture was measured at 505 nm. The result was expressed as mg/dL.

3.24. Estimation of Creatinine in Serum

The Autospan liquid gold creatinine (Jaffe's modified) reagent kit from Arkray Healthcare Pvt. Ltd. was used for the quantitative estimation of serum creatinine. The optical density of the reaction mixture was measured at 510 nm spectrophotometrically. The result was expressed as mg/dL.

3.25. Estimation of Inflamatory Markers (TNF-α and IL-6)

The estimation of TNF-α and IL-6 levels in the serum of rats treated with cartap, malathion and their mixture were performed by using the Krisgen Kits. The procedures were followed as mentioned in the kits protocol.

3.25.1. Assay Procedure for TNF-α

1) Bring all reagents to room temperature prior to use. It is strongly recommended that all standards and samples be run in duplicate or triplicate. A standard curve is required for each assay.
2) Add 100 µl/well of standards and samples to the plate. Perform two-fold serial dilutions of the 2000 pg/ml top standard, either within the plate or in separate tubes. Thus, the rat TNF-α standard concentrations are 2000 pg/ml, 1000 pg/ml, 500 pg/ml, 250 pg/ml, 125 pg/ml and 62.5 pg/ml. Assay diluent (1X) serves as the zero standard (0 pg/ml). Seal plate and incubate for 2 h at room temperature (18-25°C).
3) Aspirate and wash plate 4 times with wash buffer (1X) and blot residual buffer by firmly tapping plate upside down on absorbent paper. Wipe of any liquid from the bottom outside of the microtiter wells as any residue can interfere in the reading step. All the washes should be performed similarly.
4) Add 100 µl of diluted detection antibody solution to each well, seal plate and incubate for 1 h at room temperature (18-25°C).
5) Wash plate 4 times with wash buffer (1X) as in step 3.
6) Add 100 µl of diluted streptavidin-HRP solution to each well, seal plate and incubate for 30 min at room temperature (18-25°C).
7) Wash plate 4 times with wash buffer (1X) as in step 3. For this final wash, soak wells in wash buffer for 30 sec to 1 min for each wash. This will help minimize background.
8) Add 100 µl of TMB substrate solution and incubate in the dark for 15-30 min. Positive wells should turn bluish in color. It is not necessary to seal the plate during this step.
9) Stop reaction by adding 100 µl of stop solution to each well. Positive wells should turn from blue to yellow.
10) Read optical absorbance at 450 nm within 30 min of stopping reaction.

3.25.1.1. Calculation of Results

Determine the mean absorbance for each set of duplicate or triplicate standards and samples. Subtract the mean absorbance of the zero standards (background) from each well. Plot the standard curve on semi log graph paper, with cytokine concentration on the X-axis and absorbance on the Y-axis. Draw the best fit straight line through the standard points. To determine the unknown cytokine concentrations, find the unknowns mean absorbance value on the Y-axis and draw a horizontal line to the standard curve. At the point of intersection, draw a vertical line to the x-axis and read the cytokine concentration. If samples were diluted, multiply by the appropriate dilution factor. Computer based curve-fitting software may be preferred.

3.25.2. Assay Procedure for IL-6

1) Bring all reagents to room temperature prior to use. It is strongly recommended that all standards and samples be run in duplicate or triplicate. A standard curve is required for each assay.
2) Add 50 µl of diluted biotin conjugated detection antibody followed by 100 µl/well of standards and samples to the plate. Perform two-fold serial dilutions of the 4000 pg/ml top standard, either within the plate or in separate tubes. Thus, the rat IL-6 standard concentrations are 4000 pg/ml, 2000 pg/ml, 1000 pg/ml, 500 pg/ml, 250 pg/ml, 125 pg/ml and 62.5 pg/ml. Assay diluent serves as the zero standard (0 pg/ml). Seal plate and incubate at 37^0C for 2 h.
3) Wash plate 5 times with wash buffer (1X) and blot residual buffer by firmly tapping plate upside down on absorbent paper. Wipe of any liquid from the bottom outside of the microtiter wells as any residue can interfere in the reading step. All the washes should be performed similarly.
4) Add 100 µl of diluted streptavidin-HRP solution to each well, seal plate and incubate at 37^0C for 30 min.

5) Wash plate 5 times with wash buffer (1X) as in step 3. For this final wash, soak wells in wash buffer for 30 sec to 1 min for each wash. This will help minimize background.
6) Add 100 µl of TMB substrate solution and incubate in the dark for 15-30 min at 37^0C (do not exceed 30 min) and observe the wells every 5 min of interval. Positive wells should turn bluish in color. It is not necessary to seal the plate during this step.
7) Stop reaction by adding 100 µl of stop solution to each well. Positive wells should turn from blue to yellow.
8) Read absorbance at 450 nm within 15 min of stopping reaction.

3.25.2.1. Calculation of Results

Determine the mean absorbance for each set of duplicate or triplicate standards and samples. Subtract the mean absorbance of the zero standards (background) from each well. Plot the standard curve on semi log graph paper, with cytokine concentration on the X-axis and absorbance on the Y-axis. Draw the best fit straight line through the standard points. To determine the unknown cytokine concentrations, find the unknowns mean absorbance value on the Y-axis and draw a horizontal line to the standard curve. At the point of intersection, draw a vertical line to the x-axis and read the cytokine concentration. If samples were diluted, multiply by the appropriate dilution factor. Computer based curve-fitting software may be preferred.

3.26. Estimation of Total Protein Content

The protein content present in different samples was measured according to the method of Lowry et al. (1951) using bovine serum albumin (BSA) as a standard.

3.27. Estimation of Total DNA Content

The DNA assay was performed following the method of Burton (1956). The DNA concentration in the samples was determined by hydrolysing the

RNA in 0.1 M NaOH after which the DNA was precipitated with half volume 10% $HClO_4$. The DNA was re suspended in 10% $HClO_4$ and incubated at 70^0C to hydrolyse the DNA. After clearing the solution by centrifugation, diphenylamine and acetaldehyde were added to a final concentration of 2% and 0.01%, respectively. After an overnight incubation at 30^0C the absorbance was determined at 560 nm and 700 nm against the blank. The standard curve of absorbance at 560 nm minus absorbance at 700 nm was plotted as a function of mutual DNA concentration and then use the curve to read off unknown DNA concentratrons. The concentration in the samples was calculated relative to the included calibration curve that was prepared using highly purified calf thymus DNA.

3.28. Calculation of Protein/DNA Ratio

The ratio of protein and DNA was calculated by using the following formul:

Protein/DNA= total concentration of protein/Total DNA concentration

3.29. Calculation of Oxidative Stress Index (OSI)

The OSI has been calculated as the ratio of prooxidant/antioxidant (P/A) by using the formula described elsewhere (Jaiswal et al., 2017).

OSI = Levels of non-enzymatic antioxidant/Levels of specific activities of enzymatic antioxidants.

3.30. Histological Study of Liver Using Hematoxylin and Eosin (H & E) Staining Method

Tissues from experimental animals were quickly dissected out, washed in saline buffer, blotted and weighed. The fixation in Bouin's fixatives for overnight was performed followed by washing for 24 h under running water, dehydration in graded ethanol (30, 50, 70, 90 and 100%) and embedded in paraffin wax. The tissue embedded in paraffin was sectioned (10μm) serially using Spincon rotary microtome (model number SP-1120/SP-1120A). The sections were distributed on albumin coated sterilized glass slides, stretched and stained with H & E dyes. The staining was done by the standardized protocol for eosin-hematoxylin staining of our laboratory. The slides containing tissue sections stained with H & E were observed under a light microscope after being mounted using dibutylphthalate polystyrene xylene (DPX). All sections were evaluated for the degree of hemorrhage, vacuolar degeneration, and dilation of sinusoids, vascular congestion and necrosis. Each liver slide was examined, and the severity of the changes were scored using the scale of none (-), mild (+), moderate (++) and severe (+++).

3.31. Isozyme Profiling of AChE Using Native-PAGE (Lamelli, 1970)

3.31.1. Preparation of Rat Brain Homogenate

The rat brain was placed in an Ependauf tube and was thoroughly homogenized with the help of a micro-pestle in chilling conditions. Adding appropriate volume of buffer (0.32 M sucrose, 50 mM sodium phosphate buffer pH 7.5, 1 mM EDTA), 100% of brain homogenate was prepared. This homogenate was then centrifuged at 12000 rpm for 5 min. The debris settles and the 50 μl of the supernatant was used for further analysis. 1 mM-2.5 mM of EDTA in the homogenizing buffer provided optimum solubilization of brain AChE. To the rest of the supernatant 0.1% (v/v) of Triton X-100 was added and mixed thoroughly. It was then centrifuged at 12,000 rpm for 5 min. The debris settled in the bottomand the supernatant was used for further analysis. Triton X-100, a non-ionic detergent, is commonly used to solubilize AChE.

3.31.2. Analytical PAGE Reagents

a) Tank buffer (27.5 mM Tris, 192 mM Glycine, pH 8.3, 0.1% Triton X-100)
b) Separating gel buffer 4X (0.375 M Tris-Cl, pH 8.8, 0.1% Triton X-100, final concentration in polymerized gel)
c) Stacking gel buffer 4X (0.125 M Tris-Cl, pH 6.8, 0.1% Triton X-100, final concentration in polymerized gel)
d) Monomer solution for separating gel (40% T, 5% C)
e) Monomer solution for stacking gel (6.25% T, 20% C)
f) Catalyst (0.006% APS, 0.002% riboflavin)
g) Gel loading buffer (bromophenol blue dye): Add glycerol (3.0 ml), bromophenol blue (25 mg) and make the volume up to 10.0 ml double distilled waterand autoclaved.

3.31.3. Method

Setting up the mini electrophoresis apparatus: As mentioned above, the gel is prepared as follows:

Separating gel (7.5% T, 5% C) was prepared by mixing the monomer (3.75 ml) with the gel buffer (7.5 ml + distilled water 15.0 ml + catalyst 3.75 ml + TEMED 60 µl). Gel contents mixed well and poured in between plates, till 3/4th of the height of the plate. After the separating gel polymerizes, the stacking gel is added. The stacking gel (3.125% T, 20% C) were made by mixing monomer (10.0 ml), gel buffer (5.0 ml), distilled water, catalyst (5.0 ml) and TEMED (10.0 µl). Gel contents mixed well and poured in between plates, till the top. The comb is then inserted, taking care that no air bubbles are formed below the teeth of the comb. After the gel is polymerized, the comb is removed straight in upward direction, and care is taken not to disturb the wells. The upper and the lower tanks are filled with buffers. Samples and the tracking dye are mixed in the proportion (5:1), respectively, on a paraffin strip. With the help of a micropipette the sample is loaded in the well. The protein concentration of all the samples loaded is kept constant. The cables are connected to the mini electrophoresis unit and the power pack. The power pack is connected to a stabilizer. The current switched on, the gel is run at 15 mA for 30 min and then increased to 22.5 mA, constant current for 5 h. When the tracking dye has run $3/4^{th}$ of the gel, the power is switched off. The plates are opened with the help of a spatula and the gel is then stained for AChE activity and proteins.

3.31.4. Localization of AChE Activity

Reagents preparation for AChE staining

1) Phosphate buffer: (100 mmol/L); pH 7.0 dissolve 15.6 gm of $Na_2HPO_4.2H_2O$ in 750 ml of water. Adjust pH 7.0 by adding NaOH solution (100 mmol/L) or HCl (100 mmol/L). Check pH at 25^0C. Make up to 1.0 L with double distilled water.
2) Citrate (0.1 M): 2.941 g of trisodium citrate in 100 ml double distilled water.
3) Copper sulphate (30 mM): 0.74907g in 100 ml double distilled water.
4) Potassium-ferricyanide (5 mM): 0.1646 g in 100 ml double distilled water.
5) Solution A (freshly prepared): Citrate: $CuSO_4$: double distilled water: $K_3[Fe (CN)_6]$ [1:2:2:2].
6) Solution B = Acetylthiocholine iodide 28.92 mg in 65.0 ml phosphate buffer, 0.1 M pH 7.0
7) Incubation mixture (freshly prepared) = Solution A (35.0 ml) + Solution B (65.0 ml).

Gels were removed and placed in phosphate buffer (0.1 M, pH 7.0) for ½ h on shaker to equilibrate gel to the optimum pH for subsequent localization reaction. The gel was then transferred to incubation mix and agitated on shaker in dark for 4h until the bands were appeared. When the bands were appeared as brown zones the reaction was stoped by placing the gels in 7% acetic acid solution.

3.32. Protein Profile of Rat Brain Using SDS-PAGE (Laemmli, 1970)

3.32.1. Homogenization of the Brain

Rat brain was kept in an Ependauf and was thoroughly homogenized with the help of a micro-pestle in chilling conditions. The brain homogenate (10%, w/v) was prepared by adding appropriate volume of buffer (50mM sodium phosphate buffer pH 7.5, containing 0.2% Triton X-100). This homogenate was then centrifuged at 12000 rpm for 15 min. The debris settled at the bottom of the centrifuge tube and the the supernatant was used for further analysis.

Sodium dodecyl sulfate-polyacrylamide gel electrophoresis (SDS-PAGE) is used to separate proteins with relative molecular mass no smaller than 10 kDa. Very small proteins (<10 kDa) are difficult to resolve due to low ability of binding to SDS, which can be solved by gradient gels or using different eletrophoresis conditions, like Tricine-SDS-PAGE. The basic Laemmli SDS-PAGE procedure is described here.

3.32.2. SDS-Page

A. Making SDS-PAGE gel
   1) Clean and completely dry glass plates, combs and spacers are required.
   2) Assemble gel cassette by following manufacturer instructions.
   3) Prepare 10% lower gel (separating gel) by adding the following solutions (total volume was 5 ml): 2.0 ml double distilled water, 1.67 ml 30% acrylamide/bis, 1.25 ml of 1.5 M Tris (pH 8.8),

25.0 μl 20% SDS, 25.0 μl 10% ammonium persulfate (make it fresh and store at 4°C up to a month) and 2.5 μl TEMED (add it right before pour the gel).

4) To avoid polymerization, after adding TEMED, mix well and quickly transfer the gel solution by using 1.0 ml pipette to the casting chamber between the glass plates and fill up to about 0.7 cm below the bottom of comb when the comb is in place.
5) Add a small layer of isopropanol to the top of the gel prior to polymerization to straighten the level of the gel.
6) Once the gel has polymerized, start to prepare stacking gel (5%) by adding the following solutions (total volume was 3.0 ml): 2.08 ml double distilled water, 0.50 ml 30% acrylamide/Bis, 0.375 ml 1 M Tris (pH 6.8), 15.0 μl of 20% SDS (w/v), 15.0 μl of 10% ammonium persulfate and 1.5 μl TEMED (add it right before the gel is poured).
7) Remove the isopropanol layer by using filter paper. Rinse the top layer of the gel with double distilled waterand dry off as much of the water as possible by using filter paper.
8) Add TEMED and mix the stacking gel solution well. Quickly transfer the gel solution by using a 1.0 ml pipette till the space is full, and then insert the comb.
9) Allow the top portion to solidify and then carefully remove the comb.

B. Sample preparation

1) Prepare same amount of protein samples according to the results of protein assay.
2) Add the same volume of 2X protein sample buffer to each protein sample, mix and boil the samples at 95^0C heating block module for 10 min.
3) Spin the samples at the maximal speed for 1-2 min in tabletop centrifuge and leave the samples at room temperature until you are ready to load onto the gel.

Note: Can store extracted protein samples (containing sample buffer) at -20^{0}C and reheat at 95^{0}C for 5min when used the following time.

4) 10X running buffer was prepared by mixing 30.3 g Tris-base, 144.0 g glycine, and 10.0 g SDS. Completely dissolve in about 800 ml double distilled waterand then more double distilled waterup to 1.0 L.
5) 2X SDS protein sample buffer was prepared by mixing the 1.25 ml of 1 M Tris-HCl (pH 6.8), 4.0 ml of 10% SDS (w/v), 2.0 ml of glycerol, 0.5 ml of 0.5 M EDTA, 4 mg bromophenol blue and 0.2 ml β-mercaptoethanol (14.30 M). Bring the volume to 10.0 ml with double distilled water.

C. Electrophoresis
   1) Remove the gel cassette from the casting stand and place it in the electrode assembly with the short plate on the inside. Press down on the electrode assembly while clamping the frame to secure the electrode assembly and put the clamping frame into the electrophoresis tank.
   2) Pour some 1X electrophoresis running buffer into the opening of the casting frame between the gel cassettes. Add enough buffer to fill the wells of the gel. Fill the region outside of the frame with 1X running buffer.
   3) Slowly load the same amount of protein samples into each well and also load 10.0 μl of protein molecular weight marker.
   4) Connect the electrophoresis tank to the power supply. The running of gel was made at the voltage of 100 V for 45 min

D. Protein detection: Coomassie staining is able to detect protein bands containing about 0.2 μg or more protein. For low abundant protein, silver staining is a better choice since it is about 100-fold more sensitive than Coomassie staining. The use of silver staining is more sensitive and can detect as little as 5 ng protein. The 100.0 ml staining solution was prepared by solubilizing 0.25g of Coomassie Brilliant Blue R250 in 10.0 ml of glacial acetic acid and 90.0 ml of methanol: water (1:1, v/v). Destaining solution is the same as staining solution, but not containing the Coomassie R250 dye powder.

3.32.3. Protein Staining Procedure

1) Incubate the gel in staining solution with shaking for 30 min or longer (can leave it overnight).
2) Remove the dye solution and rinse the gel with water 1-2 times to remove the dye.
3) Add destaining solution to the gel and incubate for 30-60 min.
4) Transfer the gel to double distilled water.

3.33. Statistical Analysis

Data are presented as mean ± standard deviation using Graph Pad Prism version 5.01 for Windows, Graph Pad Software, San Diego California USA. The Graph Pad Prism version 5.01 was employed for analyzing the data using one-way analysis of variance (ANOVA) and calculating the values as mean ± standard deviation of three experiments. Bonferroni's Multiple Comparison Test was used to compare different groups. The values were considered significant at $p \leq 0.05$. The histopathological analysis of liver tissues was done using photomicrographs taken with Dewinter microscope (model-YJ-2016) attached with Dewinter microscopic camera DGI810 with Dewinter biowizzard imaging software.

Chapter 4

RESULTS

4.1. CHEMICAL CHARACTERIZATION OF AQUEOUS EXRACT OF *A. VERA* LEAVES

4.1.1. Aqueous Extraction of *A. vera* Leaves

The young leaves of *A. vera* (L.) plant (family Asphodelaceae) were collected from Roxburgh Botanical Garden, Science Faculty, Allahabad University, Prayagraj, India and freed from foreign matters by washing under running water. The plant was recognized and validated by Professor of Taxonomy, University of Allahabad. The leaves (2 kg) were dehydrated to dry (72 g) in shade at room temperature. It showed 3.6% of the dried material. This result suggested the presence of 96.4% of water content in the leaves of *A. vera*.

The shade dried leaves were powderized followed by their hot extraction (at boiling temperature) using Soxhlet apparatus in aqueous solvent (1:10 W/V; *A. vera* leaf powder and water). The Soxhlet apparatus containing 50 g of *A. vera* leaves powder with 500 ml of double distilled water was run for 24 h till the solvent eluting became colorless. After that the eluted solvent containing the *A. vera* leaves powder extract were lyophilized and stored at -20^{0}C. The total yield was 3.38 g i.e., 6.76%.

4.1.2. Chemical Properties/Phytochemical Screening of Aqueous Extract of *A. vera* Leaves

To assess the chemical properties of *A. vera* leaves powder extract the colometric assays for the estimation of different phytochemicals (flavonoids, saponins, tannins, terpinoids, steroids, phenols, quinones, carbohydrates and alkaloids) were performed using standard procedures as mentioned in Materials and Methods. The results shown in Table 1 demonstrated that the flavonoids, terpinoids, phenols andcarbohydrates were present in the *A. vera* leaves extract. However, the *A. vera* extract did not show the presence of tannins, saponins, steroids, quinones and alkaloids (Table 1).

Table 1. Phytochemical analysis of *A. vera* leaves

Plant name	Solvent	Phytochemicals	
A. vera	Aqueous	Flavonoids	Present
		Phenolic	Present
		Flavanol	Present
		Terpinoids	Present
		Carbohydrates (reducing)	Present
		Saponins	Absent
		Tannins	Absent
		Steroids	Absent
		Quinones	Absent
		Alkaloids	Absent

The phytochemical analysis of aqueous extract of *A. vera* leaves has been carried out as described in Materials and Methods section.

4.1.3. Phytochemical Evaluation of Aqueous Extract of *A. vera* Leaves

The medicinal plants have been a rich source of biologically active natural antioxidants which have been reported to play an important role in the drug discovery. The major plant-based antioxidant compounds belong to the polyphenolic group. The polyphenolic compounds are basically aromatic compounds with more than one phenolic substitution. The *A. vera* plant has been reported for its high medicinal properties and offer protection against oxidative stress mediated toxicity. Therefore, the phytochemical evaluation of aqueous extract of *A. vera* leaves was carried out for the presence of total phenolic, flavonoid and flavonol content. The results presented in Table 2 showed the presence of polyphenolic compounds.

Table 2. The composition of polyphenolic contents in aqueous extract of *A. vera* leaves

S. No.	Polyphenolic group	Concentration (mg/g of dry weight)
1	Total phenolic content	332.67±78.43 (33.26%)
2	Total flavonoids content	43.87±47.87 (4.38%)
3	Total flavanol content	13.98±13.09 (1.39%)

The determination of polyphenolic contents has been carried out as described in materials and methods section. Values are expressed as mean ±SD of 3 diffferent experiments. The values in the parenthesis indicate percent content of compounds in dry powder of *A. vera* leaves.

The aqueous extract of *A. vera* leaves were found to contain very high amount of total phenolic (332.67±78.43 i.e., 33.26%), flavonoids (43.87±47.87 i.e., 4.38%) and flavanol content (13.98±13.09 i.e., 1.39%) of the total dry weight of the *A. vera* leaves.

4.1.4. DPPH Assay to Estimate Free Radical Quenching Potential of Aqueous Extract of *A. vera* Leaves

In order to ascertain the free radical quenching potential of aqueous extract of *A. vera* leaves the DPPH assay was performed as mentioned in

Materials and Methods. The free radical scavenging potential of aqueous extract of *A. vera* leaves at different concentrations (25, 50, 100, 250, 500 and 1000 μg/ml) asdetermined by the DPPH assay showed that the extract was exhibiting free radical quenching potential in concentration dependent manner (Figure 1) i.e., the scavenging activity of DPPH radical was higher with the increasing concentration of the extract. The aqueous extracts of *A. vera* caused drastic decrease in absorbance of DPPH radical showing very high scavenging potential of the plant extract.

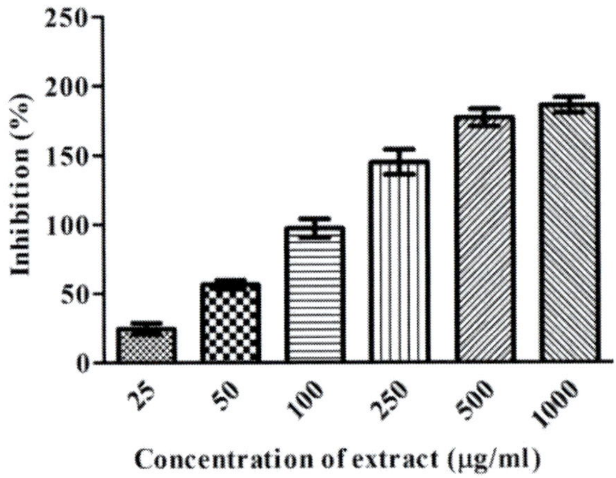

Figure 1. DPPH free radical scavenging activity of *A. vera* leaves extract. The aqueous extract was prepared and used for evaluation of its free radical scavenging activity at different concentrations of the extract (25, 50, 100, 250, 500 and 1000 μg/ml) as described in Materials and Methods. The results are expressed as mean ± SD of three replicates ($P < 0.05$).

4.2. Morphological and Behavioural Symptoms Induced by Pesticides in Rats and Effect of Aqueous Extract of *A. vera*

The morphological and behavioural alterations in rats exposed to sublethal concentration of cartap and malathion have been observed continuosly till 15 days. At the end of the treatment of pesticides upto 15 days, no death in any of the groups was observed. However, the pesticides treated groups showed appearance of varying degrees of morphometric and behavioural alterations. The symptoms included huddling and cuddling with mild tremor. When compared to control, the changes recorded in the rat body weight were found to be significant. The average relative liver weight (ARLW) and average relative brain weight (ARBW) were found to be significantly reduced in the rats treated with cartap, malathion and their combination. However, the pre-administration of *A. vera* extract to the pesticides treated groups registered an ameliorating impact on the changes in the morphological and behavioural indices (Table 3).

4.3. Effects of Pesticides and Aqueous Extract of *A. vera* Leaves on the Status of Hematological Parameters of Rat Blood

In order to assess the effect of pesticides individually and in their combination at their sublethal doses on the hematological parameters, the Wistar rats were orally administered the pesticides for 15 days at the intervals of 24h as mentioned in Materials and Methods. Rats treated with the pesticides were also pre-administered orally with aqueous extract of *A. vera* leaves for the assessment of their protective effect, if any. The results as summarized in Table 4 demonstrated that the pesticide-treated rats did not show any significantly alterations in terms of any of the hematological parameters in comparison to control. The *A. vera* extract treatment also did not show any significant change in the hematological parameters among the rats from all the experimental groups, excepting the total WBC count which increased more than two-fold in the experimental groups of rats treated with malathion and a combination of cartap and malathion (Table 4).

Table 3. Impact of cartap, malathion and *A. vera* extract on the rat body and organ (liver and brain) weight

Parameters	C*	PE+C*	C	M	C + M	PE + C	PE + M	PE+C+ M
BWI (g)	122.2± 5.41	125.2± 3.48	121.8± 1.16	122.3± 1.96	125.7± 3.01	119.3± 2.16	120.2± 2.04	116.8± 2.56
BWF (g)	131.3± 4.84	131.6± 1.47	185.3± 3.93	204.0± 4.98	211.0± 2.96	134.78± 3.50	141.05± 2.89	135.18± 3.28
BW (%)	+3.72%	+2.55%	+26.06%	+33.40%	+33.92%	+6.48%	+8.67%	+7.86%
AALW (g)	8.02 ± 0.20	8.23 ±0.15	8.56 ±0.18	8.97 ±0.30	8.84 ±0.19	8.45 ± 0.17	8.89 ±0.13	7.98 ±0.11
ARLW (g)	6.10	6.26	4.61	4.39	4.18	6.26	6.30	5.90
AABW (g)	1.914 ± 0.07001	1.951 ±0.1102	1.755 ±0.05507	1.692 ±0.1248	1.690 ±0.07376	1.970 ± 0.1012	1.937 ±0.1546	2.018 ±0.1141
ARBW (g)	1.45	1.48	0.94	0.82	0.80	1.46	1.37	1.49

Values are presented as mean ± SD of 6 rats. % body weight change/week = [Final body weight- Initial body weight/Initial body weight)/number of weeks] x 100and the Average relative organ (brain/liver) weight= [Average absolute organ weight/Final body weight] x 100 as described in Materials and Methods. The data represent mean ± SD of three independent experiments. The P values to be significant (*) at $P < 0.05$ and non-significant (ns) at $P>0.05$ as compared to control group. BWI: initial body weight; BWF: final body weight; BW (%): body weight (% change);g = grams; AALW: Average absolute liver weight; ARLW: Average relative liver weight; ARBW: Average relative brain weight; AABW: Average absolute brain weight.C*: control; PE+C*: *A. vera* treated control; C: cartap treated; M: malathion treated; C + M: cartap + malathion treated; PE+ C: cartap with pretreatment of *A. vera*; PE + M: malathion with pretreatment of *A. vera*; PE+C+M: cartap + malathion with pretreatment of *A. vera*. The data shown in bold reflect significant alterations.

Table 4. Effect of cartap, malathion and *A. vera* extract on hematological parameters of the Wistar rats in different treatment groups

HP	C*	PE+ C*	C	M	C + M	PE + C	PE + M	PE + C + M
Total WBC Count (x10³/mm³)	5.233±1.002	3.933±1.155ns	7.767±0.585 ns	12.67±1.021***	11.33±1.159***	5.63±1.514 ns	4.4±0.866 ns	6.8±0.964 ns
LYM (%)	54.93±7.242	38.43±20.6ns	75.77±5.828 ns	68.0±9.001 ns	59.4±22.69 ns	75.1±8.163 ns	44.13±6.067 ns	63.8±7.9 ns
MON (%)	15.13±2.743 ns	15.8±1.825 ns	13.0±4.244 ns	12.83±1.914 ns	11.97±5.514 ns	11.10±3.305 ns	17.77±2.715 ns	13.17±.0.923 ns
GRAN (%)	29.93±4.508 ns	45.8±18.83 ns	11.23±1.589 ns	19.17±7.821 ns	28.63±17.18 ns	13.8±4.859 ns	38.1±8.73 ns	23.03±8.822 ns
MCH (mmg)	19.63±0.9292	17.53±0.7767 ns	17.73±2.454 ns	19.23±2.914 ns	19.33±2.369 ns	19.43±0.7234 ns	19.93±0.3786 ns	21.5±0.8718 ns
MCHC (mg/dl)	40.5±0.2646	41.13±0.2062 ns	37.47±7.508 ns	38.23±6.525 ns	38.23±5.659 ns	38.6±2.858 ns	41.63±1.137 ns	40.1±0.2 ns
HGB (g/dl)	12.8±0.1 ns	12.33±0.4933 ns	13.73±0.8083 ns	10.03±3.754 ns	10.9±2.707 ns	11.83±0.6807 ns	13.5±0.1 ns	11.67±1.242 ns
MCV (μm³)	48.47±1.935	42.6±1.587 ns	47.77±3.493 ns	50.47±1.405 ns	50.63±1.358 ns	50.37±2.023 ns	47.9±2.17 ns	53.53±2.369 ns
HCT (%)	31.6±0.2	29.97±1.097 ns	36.87±5.054 ns	25.83±6.643 ns	27.8±2.787 ns	30.87±4.2 ns	32.43±0.8083 ns	29.07±2.974 ns
Total RBC Count (x10⁶/mm³)	6.527±0.2608 ns	7.033±0.1106 ns	7.697±0.4801 ns	5.12±1.309 ns	5.5±0.6872 ns	6.11±0.5747 ns	6.78±0.14 ns	5.453±0.8198 ns
PLT (x10³/mm³)	399.7±72.61	463.7±54.17 ns	584.7±11.37 ns	387.7±119.3 ns	339.7±207.6 ns	311.3±115.8 ns	248±54.62 ns	252.3±40.5 ns

Table 4. (Continued)

HP	C*	PE+ C*	C	M	C + M	PE + C	PE + M	PE + C + M
MPV (μm^3)	5.967±0.5508	5.467±0.05773 ns	5.367±0.1155 ns	5.5±0.3 ns	5.567±0.05773 ns	6.133±0.5686 ns	7.3±0.8185 ns	6.133±0.6506 ns
PCT (%)	0.2377±0.03968 ns	0.253±0.02696 ns	0.314±0.001732 ns	0.2743±0.04676 ns	0.1503±0.1032 ns	0.195±0.08479 ns	0.1973±0.05173 ns	0.1697±0.02566 ns

HP: Hematological parameters; C*:control; PE + C*: *A. vera* treated control; C: cartap treated; M: malathion treated; C + M: cartap + malathion treated; PE + C: cartap with pretreatment of *A. vera*; PE+M: malathion with pretreatment of *A. vera*; PE+C+M: cartap + malathion with pretreatment of *A. vera*; WBC: white blood cells; LYM %: percent lymphocyte; MON %: percent monocytes; GRAN %: percent granulocyte; RBC: red blood cells; HGB: hemoglobin; HCT: hematocrit; MCV: mean corpuscular volume; MCH: mean cell hemoglobin; MCHC: mean corpuscular hemoglobin concentration; PLT: platelets; MPV: mean platelet volume; PCT: plateletcrit; ns= non-significant; the data shown in bold reflect the signicant alterations as compared to control.

4.4. Effects of Pesticides and Aqueous Extract of *A. vera* Leaves on the Levels of Some Biochemical Paraameters in Rat Blood Serum

4.4.1. Effect of Pesticides and Aqueous Extract of *A. vera* Leaves on the Levels of MDA and GSH in Rat Blood Serum

To determine the effect of cartap, malathion and their combination, the male Wistar rats were exposed to these pesticides as mentioned in Materials and Methods. The pre-administration of *A. vera* extract was done through gavages as mentioned in Materials and Methods to assess its protective effect if any. The results shown in Table 5 reflected that the treatment of rats with both of these pesticides caused significant ($P < 0.05$) increase in the MDA and decrease the GSH levels in rat serum, when administered either individually or in combination. The aqueous extract of *A. vera* leaves did not exert any significant alterations in the level of MDA (3.73%) and GSH (5.30%). Each of these pesticides either used in individual (cartap/malathion) or in combination caused significant rise in the levels of MDA (i.e., 27.48, 42.52, and 25.33%, respectively) and decrease in the levels of GSH (i.e., 22.44, 26.12, 23.67%, respectively). The individual treatment with malathion exerted significantly higher level of change with increased MDA content (42.52%) and decreased GSH content (26.12%). When the rats were pretreated with the aqueous extract of *A. vera* leaves, their exposure to either cartap or malathion did not alter the levels of MDA and GSH as compared to control (Table 5).

Table 5. Impact of cartap, malathion and *A. vera* extract on the levels of MDA and GSH in rat blood serum

Parameters	C*	PE + C*	C	M	C + M	PE + C	PE + M	PE + C + M
MDA	30.24±2.00	31.37±2.24ns (+3.73%)	38.55±2.79** (+27.48%)	43.10±1.65*** (+42.52%)	37.90± 1.95** (+25.33%)	31.64±2.49ns (+4.62%)	30.23±1. 49ns (-0.03%)	33.46±1.50ns (+10.[illegible]%)
GSH	245.0±17.7	258.0±19.80ns (+5.30%)	190.0±20.80* (-22.44%)	181.0±15.10* (-26.12%)	187.00 ±10.70* (-23.67%)	224.00±18.10ns (-8.57%)	215.00±17.80ns (-12.24%)	231.0±17.60ns (-5.71%)

Values are presented as mean ± SD of 6 rats. The MDA level was expressed as nmoles ml^{-1} of MDA and the GSH was expressed as µmoles ml^{-1}. (+) and (-) sign indicates increase and decrease in the MDA and GSH contents, respectively. The data represent mean ± SD of three independent experiments. The P values significant (*) at $P < 0.05$ and non-significant (ns) at $P > 0.05$ as compared to control group.C*: control; PE+C*: *A. vera* treated control; C: cartap treated; M: malathion treated; C + M: cartap + malathion treated; PE+ C: cartap with pretreatment of *A. vera*; PE + M: malathion with pretreatment of *A. vera*; PE+C+M: cartap + malathion with pretreatment of *A. vera*. ns=non-significant.

4.4.2. Effect of Pesticides and *A. vera* Leaves Extract on the Levels of Uric Acid and Creatinine in Rat Blood Serum

In order to assess the consequence of cartap, malathion and their combination, the Wistar rats were treated as defined in Materials and Methods. The pesticide treated rats were also pretreated with the aqueous extract of *A. vera* leaves to evaluate its protecting effect, if any. The results presented in Figure 2 showed significant ($P < 0.05$) increase in the serum level of creatinine (Figure 2A) and uric acid (Figure 2B). Both pesticides used individually (cartap and malathion) or in combination caused note worthy rise in the level ofcreatinine by 62.65, 79.57 and 65.71%, and uric acid by 19.29, 30.17, 17.46% in their respective pesticide treated groups viz. cartap treated, malathion treated, cartap+malathion treated, respectively. The malathion alone recorded realtively higher impact in terms of increased levels of uric acid (30.17%) and creatinine (79.57%). The mixture of both of these pesticides did not show any significant change in the impact when compared to the effect of individual pesticides. The treatment of aqueous extract of *A. vera* leaves did not show any significant variation in the levels of creatinine (0.65%) and uric acid (1.82%). When the pesticides treated groups of animals were pretreated with the aqueous extract of *A. vera* leaves, the values of creatinine and uric acid were found to be almost near control (Figure 2).

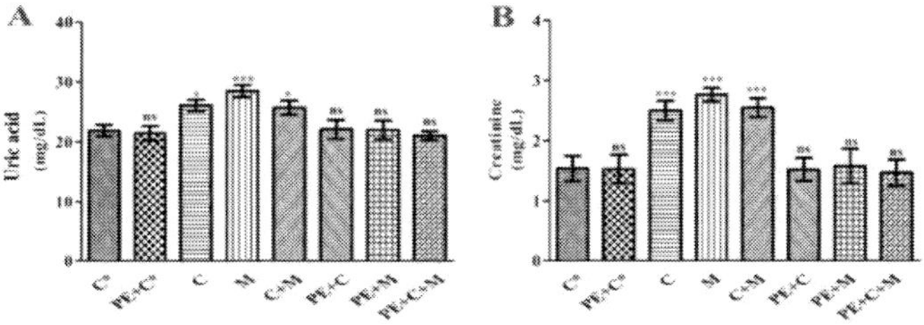

Figure 2. Effects of cartap, malathion and the aqueous extract of *A. vera* leaves on the level of (A) uric acid and (B) creatinine in rat serum. The procedures for pesticides and the aqueous extract of *A. vera* leaves administration as well as the estimation of creatinine and uric acid were followed as mentioned in Materials and Methods. The unit was expressed as mg/dL. The data represent mean ±SD of 3 independent experiments. The p values were non-significant at $p > 0.05$ as compared to control group. ns= non-significant.

4.4.3. Effects of Pesticides and Aqueous Extracts of *A. vera* Leaves on the Activities of Antioxidative Enzymes (GST, GPx, SOD and CAT) in Rat Blood Serum

The exposures of mammalian systems to different xenobiotics including peticides have shown excessive production of free radicals and generation of oxidative stress. To determine the effects of pesticides on the activities of antioxidant enzymes viz. GST, GPx, SOD and CAT, the assays for monitoring their activities were carried out using serum of rats exposed to cartap, malathion and a mixture of both as mentioned in Materials and Methods. The results suggested that the treatment of rats with these pesticides induced significant elevation in the activities of GST (Figure 3A), GPx (Figure 3B), SOD (Figure 3C) and CAT (Figure 3D).

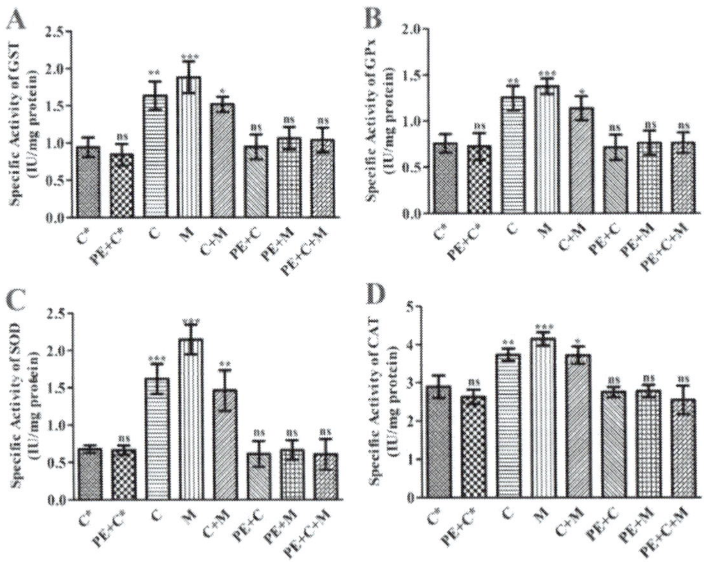

Figure 3. Effects of cartap, malathion and aqueous extract of *A. vera* leaves on the levels of antioxidant enzymes in rat blood serum. The activities of enzymes (A) GST, (B) GPx, (C) SOD, and (D) CAT were determined in the rat blood serum as mentioned in Materials and Methods section. The unit of enzyme activity was expressed as $IUmg^{-1}$ protein. The data represent mean ±SD of 3 independent experiments. The values were significant at $p< 0.0001$ as compared to control group.

The percent elevation in the activities of these enzymes was computed to be 73.82, 100.31 and 61.38% for GST; 65.19, 81.97, and 50.25% for GPx; 138.89, 216.80 and 115.58% for SOD, as well as 29.02, 43.36, and 28.54% for CAT in their respective pesticide treated groups viz. cartap treated, malathion treated, cartap+malathion treated, respectively. The mixture of both of these pesticides did not show any significant change in the impact when compared to the effect of individual pesticide. When rats were pretreated with the *A. vera* leaves extract, these groups indicated complete protection from the pesticides induced toxicity. The activities of GST, GPx, SOD, and CAT were completely protected due to pretreatment of plant extract (Figure 3). The treatment of aqueous extract of *A. vera* leaves showed almost no change in these indices even up to 15 days.

4.4.4. Effects of Pesticides and Aqueous Extract *A. vera* Leaves on the Activities of Transaminases, Phosphatases and LDH in the Rat Blood Serum

The phosphatases (ALP), transaminases (AST and ALT), and LDH are the cytosolic enzymes and considered as potential biomarkers of the status of hepatic function. The effect of sublethal doses of pesticides on the rat liver function was evaluated by estimating the activities of ALP, ALT, AST, and LDH in rat blood serum as mentioned in Materials and Methods. The protective impact of aqueous extract of *A. vera* leaf was also evaluated. The results indicated significant rise in the activities of AST by 92.10, 101.75 and 97.36% (Figure 4A), ALT by 127.61, 153.33, 131.11% (Figure 4B) and ALP by 70.07, 86.97 and 66.90% (Figure 4C) in the rat blood, respectively, in the cartap, malathion and cartap+malathion treated groups. The mixture of both of these pesticides did not show any improvement in the impact when compared to the effect of individual pesticide.

Figure 4D reflected that the treatment of rats with these pesticides either individually or in combination resulted into significant reduction in the activity of LDH by 42.20, 58.54, and 40.26%, respectively, in the cartap, malathion and cartap+malathion treated groups as compared to that of the control. When compared to the impact of individual pesticides (cartap, malathion), the effect of their combination did not show any significant reduction in LDH activity. However, the pretreatment of aqueous extract of *A. vera* leaves indicated protective effect on rats from the pesticides induced toxicity.

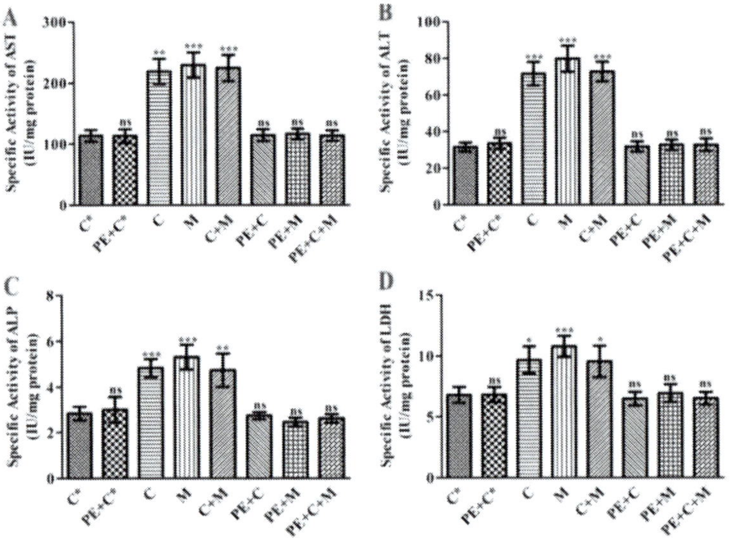

Figure 4. Effects of cartap and malathion as well as the aqueous extract of *A. vera* on the activity of transferases in rat blood serum. The activities of enzymes (A) AST, (B) ALT (C) ALP and (D) LDH were estimated in rat blood serum as mentioned in Materials and methods. The unit of each enzyme activity was expressed as $IUmg^{-1}$ protein. The data represent mean ±SD of 3 independent experiments. The values were significant at $P < 0.0001$ as compared to control group.

4.4.5. Effects of Pesticides and the Aqueous Extract of *A. vera* Leaves on the Activity of AChE from Rat Erythrocytes

Cartap and malathion are known inhibitors of AChE activity. The erythrocytes membrane contains AChE which behaves similar to that from neurotransmission system. The AChE activity is implicated as the biomarker of membrane integrity, aging, inflammation, neurotoxicity and pesticide poisoning. The effect of sublethal doses of pesticides was evaluated by assaying the activity of AChE isolated from the rat blood erythrocyte's membraneas shown in Materials and Methods. The protective impact of aqueous extract of *A. vera* leaves was also evaluated. The results indicated that the pesticides significantly ($p < 0.001$) inhibited the AChE activity by 52.84, 61.61, and 68.15%, respectively, in the cartap, malathion and their combination (Figure 5). When compared to the impact of individual pesticide, the effect of their combination did not showany more reduction in enzyme activity. The activity of erythrocyte's AChE was observed to be protected by the pre-administration of *A. vera* extract.

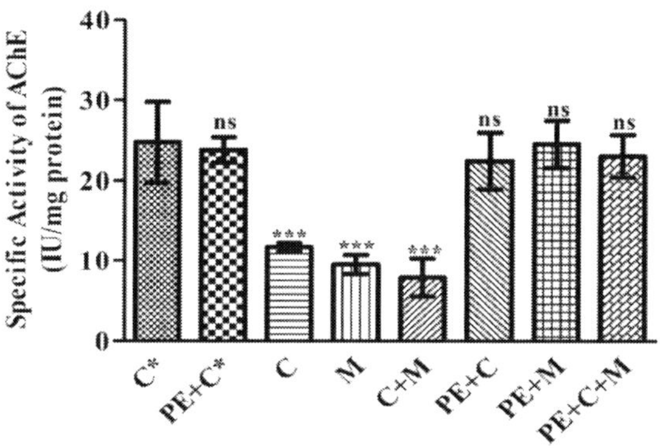

Figure 5. Effects of cartap, malathion and aqueous extract of *A. vera* leaves on the level of AChE in rat RBCs membrane as mentioned in Materials and Methods section. The unit of enzyme activity was expressed as $IUmg^{-1}$ protein. The data represent mean ±SD of 3 independent experiments. The values were significant at $p < 0.0001$ as compared to control group.

4.4.6. Effects of Pesticides and the Aqueous Extract of *A. vera* Leaves on the Levels of TNF-α and IL-6 in Rat Blood Serum

Cytokines are the small proteins or peptides functions as immunomodulating agents and also involved in autocrine, paracrine and endocrine signaling. The cytokines include tumor necrosis factors, chemokines, interleukins, lymphokines and interferons. They play animportant role in the health and disease such as theinflamation, trauma, immune response and cancer. The serum level of proinflamatory cellular cytokines is considered as the potential markers for inflammation and cellular damage. The concentrations of TNF-α and IL-6 in the blood serum of rats exposed to sublethal doses of pesticides are shown in Figure 6. The pro-inflammatorycytokines levels were increased significantly ($p<0.05$) in the rats exposed to cartap, malathion and their combination. The concentrations of TNF-α and IL-6 in the serum were increased by 85.44, 163.76, and 80.77% for TNF-α as well as 53.86, 116.15, and 83.37% for IL-6, respectively, in the experimental groups treated with cartap, malathion and their combination. When compared to the impact of individual pesticide, the effect of their combination did not showany more elevation in the levels of proinflamatory markers. The levels of these proinflamatory markers were observed to be protected by the pre-administration of *A. vera* extract.

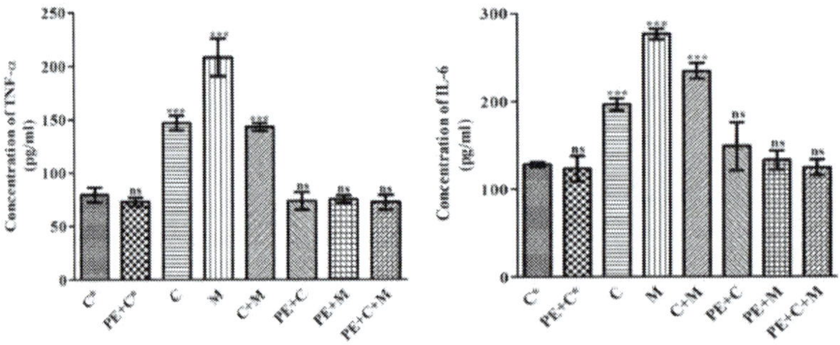

Figure 6. Effects of cartap, malathion and the aqueous extract of *A. vera* leaves on the levels of proinflamatory markers (TNF-α and IL-6) in the rat blood serum. The procedures for the administration of pesticides, aqueous extract of *A. vera* leaves and the assay for TNF-α and IL-6 were same as mentioned in Materials and Methods. The unit of the concentration of TNF-α and IL-6 were expressed as $pgml^{-1}$. The data represent mean ±SD of 3 independent experiments. The values were significant at $p< 0.05$ as compared to control group. ns= non-significant.

4.5. EFFECTS OF PESTICIDES AND AQUEOUS EXTRACT OF *A. VERA* LEAVES ON THE LEVELS OF SOME BIOCHEMICAL PARAMETERS IN THE BRAIN OF WISTAR RATS

4.5.1. Effects of Pesticides and Aqueous Extract of *A. vera* Leaves on the Levels of Malonaldialdehyde (MDA) and Glutathione (GSH) in Rat Brain

In the above experiments we observed that biochemical changes in the rat erythrocyte membrane due to exposure of sublethal concentrations of cartap and malathion, which suggests that the pesticides through blood circulation system would also be reaching to the brain of rats and causing perturbations in its oxidative status. In order to evaluate the impact of cartap, malathion and a combination of both of these pesticides at their sublethal doses, the Wistar rats were treated for 15 days at the intervals of 24h as described in Materials and Methods. The treated rats were also preadministered orally with aqueous extract of *A. vera* leaves to assess its protective effect, if any. The results presented in Table 6 indicated significant ($p < 0.05$) increase in the levels of MDA and decrease in the level of GSH in the rat brain when compared to that of the control.

The pesticides (cartap and malathion) have caused significant increase in the level of MDA by i.e., 81.22, 102.88, and 80.29% and marked decrease in the level of GSH by i.e., 40.92, 57.08, 48.52%, respectively, in the cartap, malathion and their combination. The effect of malathion individually registered relatively higher level of alterations with the increased levels of MDA (102.88%) and decreased levels of GSH (57.08%). When compared to the impact of individual pesticides (cartap and malathion), the effect of their combination did not show any significant change in the levels of MDA and GSH. When the pesticide treated animals were subjected to the pretreatment of the aqueous extract of *A. vera* leaves, the values of MDA and GSH was found to be almost the same as that of the control (Table 6). The treatment of aqueous extract of *A. vera* leaves itself did not show any significant change in the levels of these indices (3.34% for MDA and 1.56% for GSH) in rat brain.

4.5.2. Effects of Pesticides and Aqueous Extract of *A. vera* Leaves on the Activities of GST, GPx, SOD and CAT in the Rat Brain

The exposures to various xenobiotics including peticides have shown excessive production of free radicals and generation of oxidative stress in mammalian systems. To evaluate the effects of pesticides on the activities of antioxidant enzymes viz. GST, GPx, SOD and CAT, they were assayed using the homogenates of rat brain treated with cartap, malathion and a combination of both for 15 days at the intervals of 24 h as mentioned in Materials and Methods. The protective impact of aqueous extract of *A. vera* leaves was assessed by preadministering the plant extract to the pesticide treated groups. The results suggested that the treatment of rats with these pesticides showed significant decrease in the activities of GST (Figure 7A), SOD (Figure 7C) and CAT (Figure 7D) except the activity of GPx (Figure 7B) which was observed to be increased signicicantly as compared to the control. The percent alteration in the activities of these enzymes was computed to be 30.53, 59.96 and 29.87% for GST,35.62, 46.84, and 39.27% for GPx,25.94, 35.26 and 22.16% for SOD, and 26.29, 39.07, and 27.90% for CAT in their respective pesticide treated groups viz. cartap treated, malathion treated and cartap+malathion treated, respectively. The mixture of both of these pesticides did not show any further change in the impact when compared to the effect of individual pesticide. When rats were pretreated with the *A. vera* leaves extract, these groups indicated complete protection from the pesticides induced toxicity. The activities of GST, GPx, SOD and CAT were completely protected due to pretreatment of plant extract (Figures 7). The treatment of aqueous extract of *A. vera* leaves showed almost no change in these indices even up to 15 days.

Table 6. Effects of cartap and malathion as well as aqueous extract of *A. vera* leaves on the levels of malondialdehyde (MDA) and reduced glutathione (GSH) in rat brain

Parameters	C*	PE + C*	C	M	C + M	PE + C	PE + M	PE + C + M
MDA	10.76±0.6684	11.12 ±0.7865 (+3.345%)	19.50±1.735*** (+81.226%)	21.83±2.074*** (+102.881%)	19.40± 0.9535*** (+80.297%)	11.45±1.191 ns (+6.412%)	11.85±0.5784 ns (+10.130%)	11.87±1.405 ns (+10.315%)
GSH	2.104±0.244	2.137±0.2245 (+1.568%)	1.243±0.0693*** (-40.922%)	0.9029±0.0908*** (-57.086%)	1.083±0.09322*** (-48.526%)	2.164±0.2492 ns (+2.851%)	1.99±0.1955 ns (-5.418%)	2.197±0.2012 ns (+4.420%)

The procedure for the administration of pesticides and aqueous extract of *A. vera* leavesas well as the evaluation of MDA and GSH level in rat brain was same as mentioned in Materials and Methods. The MDA level was expressed as nmoles of MDA mg^{-1} protein. The GSH was expressed as $\mu g mg^{-1}$ protein. (+) and (-) sign shows increase and decrease in the level of MDA and GSH, respectively. The data represent mean ± SD of three independent experiments. The p values were significant (*) at $p < 0.05$ and non-significant (ns) at $p>0.05$ as compared to control group. C*: control; PE +C*: *A. vera* treated control; C: cartap treated; M: malathion treated; C + M: cartap + malathion treated; PE + C: cartap with pretreatment of *A. vera*; PE + M: malathion with pretreatment of *A. vera*; PE + C + M: cartap + malathion with pretreatment of *A. vera*. The highlighted figures show marked alterations. ns= non-signicant.

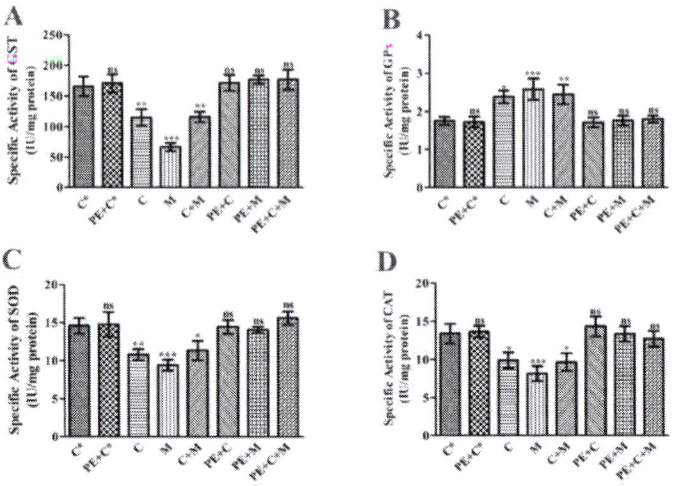

Figure 7. Effects of cartap, malathion and aqueous extract of *A. vera* leaves on the levels of antioxidant enzymes in rat brain. The activities of enzymes (A) GST, (B) GPx, (C) SOD, and (D) CATwere determined in the rat brain as mentioned in Materials and Methods. The unit of enzyme activity was expressed as $IUmg^{-1}$ protein. The data represent as mean ±SD of 3 independent experiments. The values were significant at $p < 0.0001$ as compared to control group.

4.5.3. Effects of Pesticides and Aqueous Extract of *A. vera* Leaves on the Activity of LDH in the Rat Brain

The LDH is a key cytosolic enzyme of energy metabolism and may function as the indicator of cellular damage. The elevation in the level of this marker enzyme is apparently due to the cytoskeletal changes in the brain cell caused due to xenobiotic exposure. To evaluatethe effects of pesticides on the activity of LDH, the assays for monitoring the activity of enzyme was carried out using the homogenates of rat's brain treated with cartap, malathion and a combination of both, for 15 days at the intervals of 24h as mentioned in Materials and Methods. The protective impact of aqueous extract of *A. vera* leaves was assessed by preadministering the plant extract to the pesticide treated groups. The results suggested that the treatment of rats with these pesticides showed significant decrease in the activity of LDH (Figure 7) as compared to the control. The percent decrease in the activity of this enzyme was computed to be 36.16, 55.35, and 39.71%, respectively, in their respective pesticide treated groups viz. cartap, malathion and cartap + malathion treated rats. The mixture of both of these pesticides did not show any significant change in the impact when compared to the effect of individual pesticide. When rats were pretreated with the *A. vera* leaves extract, these groups indicated complete protection from the pesticides induced toxicity. The activity of LDH was completely protected due to the pretreatment of plant extract (Figures 8). The treatment of aqueous extract of *A. vera* leaves showed almost no change in these indices even up to 15 days.

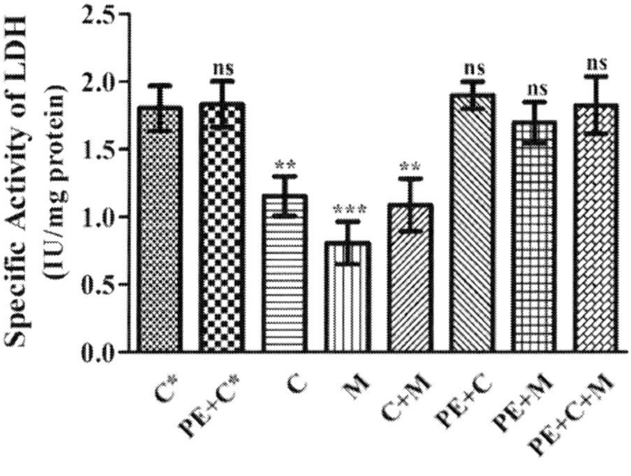

Figure 8. Effects of cartap, malathion and the aqueous extract of *A. vera* leaves on the activity of LDH in rat brain. The procedures for the administration of pesticides, aqueous extract of *A. vera* leaves and the activity assay of LDH were same as mentioned in Materials and Methods. The unit of enzyme activity was expressed as $IUmg^{-1}$ protein. The data represented as mean ±SD of 3 independent experiments. The values were significant at $p < 0.001$ as compared to control group.

4.5.4. Effects of Pesticides and Aqueous Extract of *A. vera* Leaves on the Activity of AChE in the Rat Brain

Cartap and malathion are known inhibitors of AChE activity. The rat brain contains AChE which plays a keyrole in the neurotransmission system.

To assess the effects of sublethal doses of pesticides (individually and in combination of cartap and malathion) on the level of AChE activity was evaluated by assaying the activity of AChE isolated from the rat brain homogenateas mentioned in Materials and Methods. The protective impact of aqueous extract of *A. vera* leaves was also evaluated. The results indicated that the pesticides significantly ($p < 0.001$) inhibited the AChE activity by 29.42, 38.96, and 36.54%, respectively, in the cartap, malathion and their combination (Figure 9). When compared to the impact of individual pesticide, the effect of their combination did not show marked reduction in the enzyme activity. Theactivityof AChE extracted from rat brainwas observed to be protected by the pre-administration of aqueous extract of *A. vera* leaves. However, the pretreatment of aqueous extract of *A. vera* leaves alone did not show any significant alteration in the level of AchE activity even after treatment for 15 days.

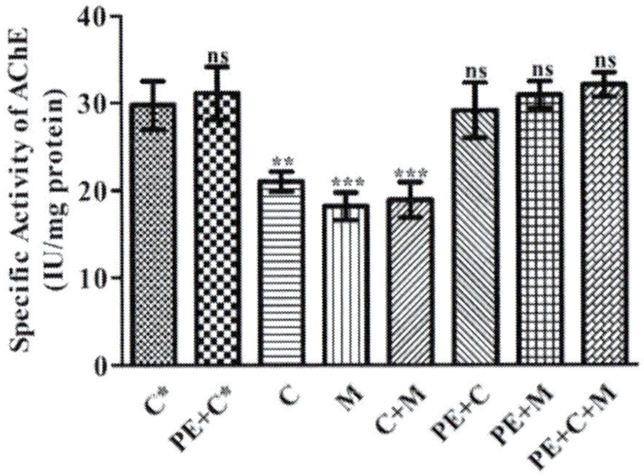

Figure 9. Effects of cartap, malathion and the aqueous extract of *A. vera* leaves on the activity of AChE in rat brain. The procedures for the administration of pesticides, aqueous extract of *A. vera* leaves and the assay for AChE activity were same as mentioned in Materials and Methods. The unit of enzyme activity was expressed as $IUmg^{-1}$ protein. The data represent as mean ±SD of 3 independent experiments. The values were significant at $p < 0.001$ as compared to control group. ns= non-significant.

4.5.5. Effects of Pesticides and Aqueous Extract of *A. vera* Leaves on the Activity of AChE Isozymes in the Rat Brain

AChE are proteins that are defined by their ability to catalyze the hydrolysis of ester bonds within lipophilic compounds. Since AChE is reported to exist in four different isozymic forms, it is expected that these pesticides may exhert differential impact on their activities. To ascertain the impact of sublethal doses of pesticides (individually and in combination of cartap and malathion) on the status of rat brain AChE isozymes, the isozyme activity profile was resolved in 7.5% Native-PAGE gel as described in the Materials and Methods. The PAGE analysis of AChE preparation showed a smear of bands suggesting the presence of four isozymes in rat brain treated with the cartap, malathion and a combination of both the pesticides. However, there was not any significant changeobserved in the isozyme activity of AChE extracted from the rat brain treated with the cartap, malathion and their combination (Figure 10). The pretreatment of aqueous extract of *A. vera* leaves also did not showed any disturbance in the AChE isozyme activity in the pesticide treated animals as well as the alone itself.

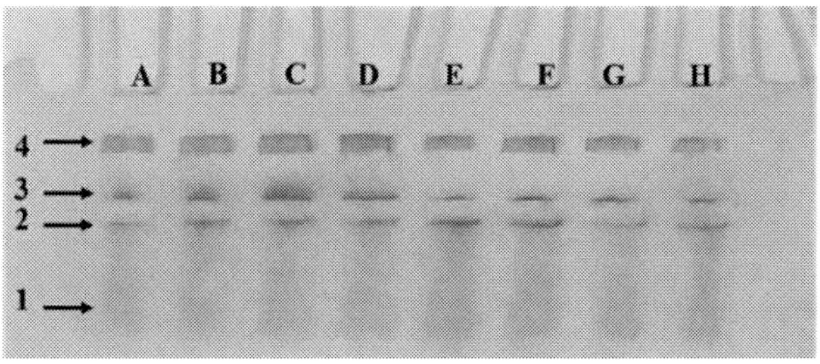

Figure 10. Effects of cartap, malathion and the aqueous extract of *A. vera* leaves on the activity of AChE isozymes of rat brain. The procedures for the administration of pesticides, aqueous extract of *A. vera* leaves and the activity assay for AChE on the gel were same as mentioned in Materials and Methods. Abbreviations: A- Control, B- Control treated with aqueous extract of *A. vera*, C- Cartap treated, D- Malathion treated, E- Cartap + Malathion treated, F- Cartap treatment with the pretreatment of aqueous extract of *A. vera*, G- Malathion treatment with the pretreatment of aqueous extract of *A. vera*, H- Cartap+Malathion treatment with the pretreatment of aqueous extract of *A. vera*. The presence and location of four different isozymes of AChE has been shown.

4.5.6. Effects of Pesticides and Aqueous Extract of *A. vera* Leaves on the Levels of Protein Profile Using SDS-PAGE in Rat Brain

To evaluate the impacts of cartap, malathion and a combination of both pesticides on the levelof rat brain protein profile at their sublethal doses was determined by performing the SDS-PAGE as described in Materials and Methods. The protective effect of pretreatment of rats with aqueous extract of *A. vera* leaves was also evaluated. The data presented in Figure 11 indicated that pesticide treatment caused variations in the intensity as well as the content of the protein bands in the cartap, malathion and a combination of cartap and malathion treated groups, respectively, as compared to the control. The combination of these two pesticides also did not showed any further changes in the protein profile when compared to the individual pesticide treated groups.

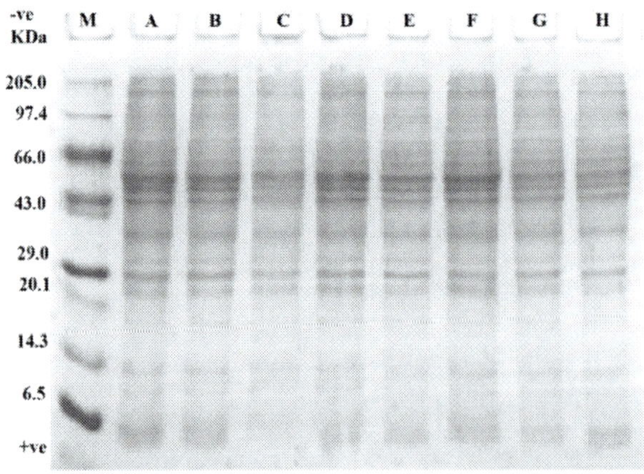

Figure 11. Effects of cartap, malathion and the aqueous extract of *A. vera* leaves on the protein profile of rat brain using SDS-PAGE. The procedures for the administration of pesticides, aqueous extract of *A. vera* leaves and the PAGE were same as mentioned in Materials and Methods. Abbreviations: A- Control, B- Control treated with aqueous extract of *A. vera*, C- Cartap treated, D- Malathion treated, E- Cartap + Malathion treated, F- Cartap treatment with the pretreatment of aqueous extract of *A. vera*, G- Malathion treatment with the pretreatment of aqueous extract of *A. vera*, H- Cartap + Malathion treatment with the pretreatment of aqueous extract of *A. vera*, kDa- Kilo Dalton, -ve- Negative, +ve- Positive.

Table 7. Protein analysis of SDS-PAGE in the brain of rats treated with cartap, malathion and the aqueous extract of *A. vera* leaves

S. No	M	A	B	C	D	E	F	G	H
1	205.0	250.0	250.0	241.0	241.0	250.0	250.0	250.0	250.0
2	97.4	213.4	216.1	210.9	210.9	210.9	213.4	213.4	213.4
3	66.0	167.4	171.5	167.4	165.3	163.3	163.3	165.3	167.4
4	46.0	150.0	151.8	151.8	150.0	147.6	148.8	150.0	151.8
5	29.0	140.6	140.6				140.6	142.9	144.0
6	20.1	132.8	135.0	132.8	132.8	131.7	131.7	133.9	135.0
7	14.3	112.0	115.7	114.8	113.9	112.9	113.9	115.7	117.6
8	6.5	102.5	102.5	105.0	102.5	100.8	102.5	103.3	105.8
9		95.0	95.0	96.5	95.0	94.5	94.5	95.5	96.5
10		86.2	86.2	87.0	86.2	85.7	85.7	87.0	87.9
11		81.0	80.6	81.4	81.0	80.2	80.6	81.0	82.3
12		72.9	72.9	75.8	75.0	72.9	72.9	75.4	76.6
13		44.4	44.3	44.6	44.6	44.3	44.3	44.6	45.2
14		38.9	39.0	39.2	39.2	38.9	38.9	39.0	39.3
15		34.3	34.7	35.4	35.4	34.7	34.3	35.0	35.4
16		27.9	27.9	29.1	28.8	28.2	28.2	28.5	29.1
17		24.0	24.2	24.4	24.4	24.2	24.0	24.2	24.2
18		23.0	22.9	23.1 22.1 21.0	23.0 22.2	23.0 21.9	22.9	23.0	23.0
19		18.7	18.7	18.9	18.9	18.7	18.7	18.7	18.7
20		16.6	16.7	17.0	17.0	16.9	16.9	16.8	16.7
21		15.0	15.0		15.0	15.0	15.0	15.0	15.0
22		15.0	15.0				15.0	15.0	15.0

Abbreviations: M- Marker, A- Control, B- Control treated with aqueous extract of *A. vera*, C- Cartap treated, D- Malathion treated, E- Cartap + Malathion treated, F- Cartap treatment with the pretreatment of aqueous extract of *A. vera*, G- Malathion treatment with the pretreatment of aqueous extract of *A. vera*, H- Cartap + Malathion treatment with the pretreatment of aqueous extract of *A. vera*.

The pretreatment of aqueous extract of *A. vera* leaves alone did not show any significant alteration in the level of protein profile in rat brain even after 15 days oftreatment. Comparing the treatments with its corresponding control, several expressed protein bands were visualized after exposure to the cartap, malation and a combination of both in the rat's brain. However, the protein of molecular weight of 140 kDa was missing in all the treatment groups and 15 kDa protein in the cartap treated groups. The intensity of the

15 kDa protein was observed to be decreased in malathion and a combination of cartap and malathion. The summary of the PAGE results has been summarized in Table 7. In addition to several up- and down-regulated proteins, new protein bands were appeared when compared to the corresponding control. SDS protein profile showed 22 common bands in control and all treatments, although there were changes in bands intensity. Also, it could be observed that the treatment with these pesticides at sublethal concentration decreased bands number, this was constant with the results of total protein content which revealed the decrease compared with the control group.

4.6. Effects of Pesticides on the levels of Some Biochemical and Histological Paraameters in the Liver of Wistar Rats; Amelioration by Aqueous Extract of *A. vera* Leaves

The reports from some workers have indicated that the exposure of rats and other mammalian systems have evoked the oxidative stress in the brain as well as in the liver causing both an increase in free radical species and decrease in antioxidant defence ability. However, the information regarding the effect of exposure of mixture of pesticides is lacking. Keeping these facts in view we have endevoured to study the effect of cartap, malathion and their combination on different cellular and physiological indices of Wistar rats.

4.6.1. Effects of Pesticides on the Levels of Non-enzymatic Antioxidants (MDA and GSH) in Rat Liver and the Ameliorative Impact of Aqueous Extract of *A. vera* Leaves

In order to evaluate the impact of cartap, malathion and a combination of both of these pesticides at their sublethal doses, the Wistar rats were treated for 15 days at the intervals of 24h as described in Materials and Methods. The treated rats were also administered orally with aqueous extract of *A. vera* leaves to assess its protective effect, if any. The results presented in Table 8 indicated significant ($p < 0.05$) increase in the levels of MDA and decrease in the level of GSH in the rat liver tissue when compared to that of the control. The pesticides (cartap and malathion) have caused significant increase in the level of MDA by 25.197, 27.964 and 63.537%, and decrease in GSH level by 35.47, 43.139 and 39.409%, respectively, in the cartap, malathion and their combination. The effect of combination of these pesticides registered relatively higher level of alterations with the increased values of MDA (63.537%) and decreased level of GSH (39.409%). When compared to the impact of individual pesticides (cartap and malathion), the effect of their combination did not show any further change in the levels of MDA and GSH. However, when the pesticide treated animals were subjected to the pretreatment of the aqueous extract of *A. vera* leaves, the values of MDA and GSH were found to be almost the same as that of the control (Table 8). The treatment of aqueous extract of *A. vera* leaves itself did not show any significant change in the levels of these indices (5.583 and 8.866%, respectively) even after 15 days of treatment.

4.6.2. Effects of Pesticides and Aqueous Extract of *A. vera* Leaves on the Activities of Antioxidant Enzymes (SOD, CAT and GST) in the Rat Liver

The peticides exposures to mammals have shown excessive production of free radicals and generation of oxidative stressin the mammalian systems. To evaluatethe effects of pesticides on the activities of antioxidant enzymes viz. SOD, CAT and GST, the assays for monitoring the activities of these enzymes were carried out using the homogenates of rat liver treated with cartap, malathion and a combination of both for 15 days at the intervals of 24h as mentioned in Materials and Methods. The protective effect of aqueous extract of *A. vera* leaves was estimated by preadministering the plant extract to the pesticides treated groups. The results suggested that the treatment of rats with these pesticides showed significant reduction in the activities of SOD (Figure 12A), CAT (Figure 12B) and GST (Figure 12C) as compared to the control.

Table 8. Effects of cartap, malathion and *A. vera* extract on the levels of malondialdehyde (MDA), and reduced glutathione (GSH) in rat liver

Parameter	C*	PE+ C*	C	M	C+M	PE+C	PE+ M	PE+ C+ M
MDA	20.24±1.790	21.37±2.008 (+5.583%)	25.34±2.492** (+25.197%)	25.90±2.818*** (+27.964%)	33.10± 1.482*** (+63.537%)	21.64±2.229 ns (+6.910%)	20.23±1.341 ns (-0.05%)	23.46±1.345 ns (+15.909%)
GSH	1.421±0.102	1.547±0.082 (-8.866%)	0.917±0.127*** (-35.47%)	0.808±0.098*** (-43.139%)	0.861±0.068*** (-39.409%)	1.149±0.071 ns (-19.142%)	1.138±0.132 ns (-19.916%)	1.207±0.110 ns (-15.08%)

The procedure for the administration of pesticides and aqueous extract of *A. vera* leaves as well as evaluation of level of MDA and GSH in rat liver was the same as mentioned in Materials and Methods. The unit of MDA level was expressed as nmoles of MDA mg^{-1} protein and the GSH was expressed as $\mu g mg^{-1}$ protein in rat liver tissues. (+) and (-) sign shows increase and decrease, respectively, in the MDA and GSH contents. The data represent mean ± SD of three independent experiments. The p values are significant (*) at $p < 0.05$ and non-significant (ns) at $p>0.05$ as compared to control group.

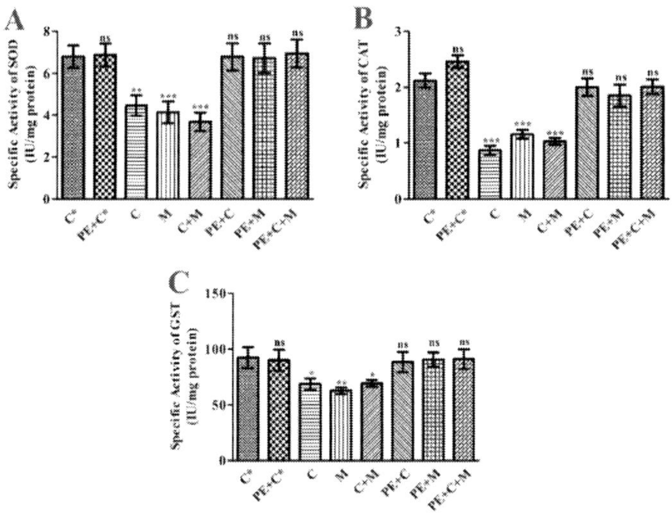

Figure 12. Effects of cartap, malathion and aqueous extract of *A. vera* leaves on the levels of antioxidant enzymes in rat liver. The activities of enzymes (A) SOD, (B) CAT and (C) GST were determined in the rat liver as mentioned in Materials and Methods. The unit of enzyme activity was expressed as $IUmg^{-1}$ protein. The data represent mean ±SD of 3 independent experiments. The values were significant at $p< 0.05$ and non-significant at $p>0.05$ as compared to control group.

The percent alteration in the activities of these enzymes was calculated to be 34.53, 39.24 and 46.01% for SOD, 59.00, 45.75, and 51.46% for CAT, and 25.49, 31.52, and 24.41% for GST in their respective pesticide treated groups viz. cartap, malathion and cartap + malathion treated, respectively. The mixture of both of these pesticides did not show any significant change in the impact when compared to the effect of individual pesticide. When rats were pretreated with the aqueous extract *A. vera* leaves, these groups displayed complete protection of rat liver from the toxic effect of pesticides. The activities of SOD, CAT and GST were completely protected due to pretreatment of aqueous extract *A. vera* leaves (Figures 12). The treatment of aqueous extract of *A. vera* leaves showed almost nonsignificant change in these indices (SOD: 1.17%; CAT: 16.03%; and GST: 2.46%) even up to 15 days.

4.6.3. Effects of Pesticides and Aqueous Extract *A. vera* Leaves on the Activities of Transaminases and Phosphatases in the Rat Liver

The transaminases (AST and ALT) and phosphatases (ACP and ALP) are the cytosolic enzymes and considered as potential biomarkers of the status of hepatic function. The effect of sublethal doses of pesticides on the rat liver function was evaluated by monitoring the activities of AST, ALT, ACP and ALP in rat liver as described in Materials and Methods. The protective effect of aqueous extract of *A. vera* leaves was also evaluated by preadministering the plant extract to these pesticides treated groups. The results indicated significant increase in the activities of AST by 38.57, 47.67 and 31.76% (Figure 13A) and ALT by 25.91, 35.10, 24.08% (Figure 13B) and ALP by 31.92, 42.77 and 30.72% (Figure 13C) and significant decrease in the levels of ACP by 29.82, 32.10, and 28.94% (Figure 13D) in the rat liver, respectively, in the cartap, malathion and cartap+malathion treated groups. The mixture of both of these pesticides did not show any significant change in the impact when compared to the effect of individual pesticide. When rats were pretreated with the aqueous extract *A. vera* leaves, these groups displayed complete protection of rat liver function from the hepatotoxic effect of pesticides. The activities of AST, ALT, ACP and ALP were completely protected due to pretreatment of aqueous extract *A. vera* leaves (Figure 13).

4.6.4. Effects of Pesticides and Aqueous Extracts of *A. vera* Leaves on the Activities of Lactate Dehydrogenase (LDH) in the Rat Liver

The LDH is a key cytosolic enzyme of energy metabolism and may function as the the indicator cellular damage and a marker for the energy crisis in the cell. To evaluatethe effects of pesticides on the activity of LDH, the assay for monitoring the activity of enzyme was carried out using the homogenates of rat's liver tissue isolated from rat reated with cartap, malathion and a combination of both for 15 days at the intervals of 24 h as mentioned in Materials and Methods. The protective impact of aqueous extract of *A. vera* extract was assessed by preadministering the plant extract to these pesticides treated groups. The results suggested that the treatment of rats with these pesticides cased a significant reduction in the LDH activity (Figure 14) as compared to the control. The percent decrease in the activity of this enzymes was computed to be 26.44, 33.59, and 33.00%, respectively, in their respective pesticide treated groups viz. cartap, malathion and cartap+malathion treated, respectively. The mixture of both of these pesticides did not show any significant change in the impact when compared to the effect of individual pesticide. When rats were pretreated with the *A. vera* leaves extract, these groups indicated complete protection from the pesticides induced toxicity. The activity of LDH was completely restored to the normal level due to pretreatment of plant extract (Figures 14). The treatment of aqueous extract of *A. vera* leaves showed almost no change in this parameter even up to 15 days.

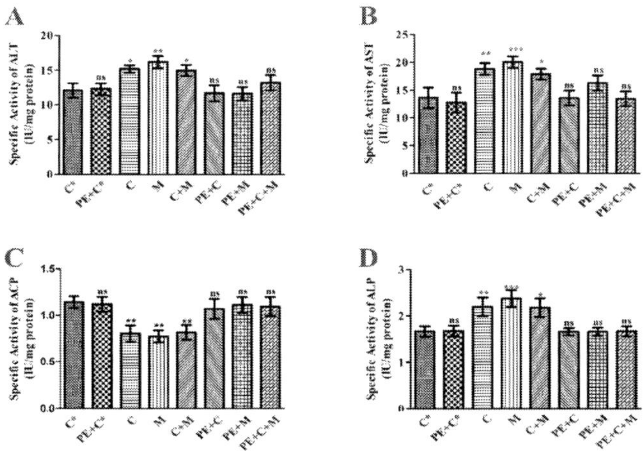

Figure 13. Effects of cartap, malathion and the aqueous extract of *A. vera* leaves on the activity of transferases in rat blood serum. The activities of enzymes (A) ALT, (B) AST (C) ACP and (D) ALP were estimated in rat liver as mentioned in Materials and Methods. The unit of each enzyme activity was expressed as IUmg^{-1} protein. The data represent mean ±SD of 3 independent experiments. The values were significant at $p< 0.05$ and non-significant at $p>0.05$ as compared to control group.

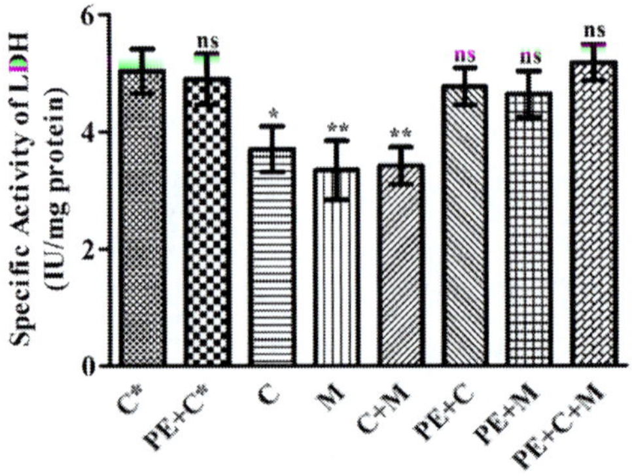

Figure 14. Effects of cartap, malathion and the aqueous extract of *A. vera* leaves on the activity of LDH in rat liver. The procedures for the administration of pesticides, aqueous extract of *A. vera* leaves and the activity assay of LDH were same as mentioned in Materials and Methods. The unit of enzyme activity was expressed as $IUmg^{-1}$ protein. The data represented as mean ±SD of 3 independent experiments. The values were significant at $p< 0.05$as compared to control group.

4.6.5. Effects of Pesticides and Aqueous Extract of *A. vera* Leaves on the Histological Parameters of Rat Liver

The xenobiotics including pesticides are known to cause the serious histological aberations in the key organs of the mammalians systems. To examine the effects of pesticides on the levels of histological parameters in the rat liver tissues, the 15 days of treatment of both the pesticides (cartap and malathion,) and their combination were processed and analyzed as mentioned in Materials and Methods. The microscopic examination of the tissue slides as presented in the Figures 15A and 15B indicated the normal histological architecture of liver to contain acini and hexagonal lobules. The hexagonal lobules were centered on central vein (CV) and had portal triad containing hepatic artery (HA), portal vein (PV) and bile duct (BD). The data revealed the appearance of mild disorganization of the CV and hepatic cords in the rat liver tissues treated with cartap (Figures 15C), malathion (Figure 15D) and their combination (Figure 15E). Liver sections from the pretreatment of rats with aqueous extract of *A. vera* leaves exhibited moderate recovery/protection (Figures 15F, 15G, and 15H) from the pesticide induced damage. The scoring of the histopathological changes in rat liver has been summarized in Table 9.

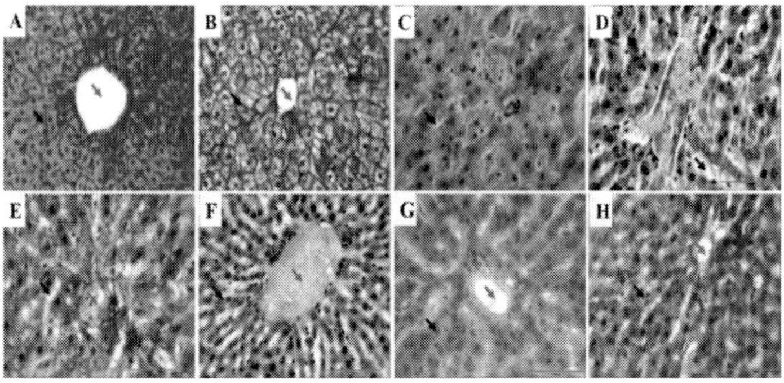

Figure 15. Histological alterations in the liver at 15 days after administration of cartap and malathion as well as the *A. vera* extract (H & E staining, 40X). (A) Photomicrograph of a liver in the control group. (B) control group treated with *A. vera* extract (positive control group), (C) cartap treated group (10% LD_{50}), (D) malathion treated group (10% LD_{50}), (E) cartap+malathion treated group (5% LD_{50} of each), (F) cartap treated group with the pre-administration of *A. vera* extract, (G) malathion treated group with the pre-administration of *A. vera* extract, (H) cartap+malathion treated group with the pre-administration of *A. vera* extract. The treatment regimen for both the pesticides and *A. vera* extracts was as described in Materials and Methods sections. The blue arrow indicates congestion in central vein (CV) and the black arrow indicates the disorganization of hepatic cords (HC).

Table 9. Effects of cartap, malathion and *A. vera* extract on the levels of architectural damage on histopathological examination of the rat liver in the different treatment groups

Parameters	C*	PE+ C*	C	M	C+M	PE+C	PE+ M	PE+ C + M
DH	-	-	++	+++	+++	-	+	+
CNH	-	-	+++	++	++	_	+	+
CCV	-	-	++++	++	+++	+	+	+
DHT	-	-	++++	+++++	++++	+	+	++

DH: degenerations of hepatocytes; CNH: coagulative necrosis in hepatocytes; CCV: congestion in central vein; DHT: degeneration in hepatic triad; (-): absent; (+): present.

4.7. Effects of Pesticides and Aqueous Extracts of *A. vera* Leaves on the Levels of Total Protein Content, Total DNA Content and Their Ratio (Protein/DNA) in Blood Serum, Packed Red Blood Corpuscles, Brain and Liver of Wistar Rat

The previous findings reported that the rats treated with cartap, malathion and their combination have strogly affected the biochemical and morphological i.e., weight of oran and body, parameters of Wistar rats. To ascertain this, we endevoured to evaluate the effects of cartap, malathion and a combination of both pesticides on the levels of total protein content in rat blood (serum and PRBC), brain and liver of Wistar rats. In order to evaluate the effect of these pesticides and the aqueous extract of *A. vera* leaves on the level of total protein content, the procedures for the administration of pesticides and the preadministration of aqueous extract of *A. vera* leaves were the same as mentioned in Materials and Methods section. The results from present study indicated that the total protein content in serum (Figure 15A), brain (Figure 15C and 15D) and liver (Figure 15E) was significantly decrased whereas in case of PRBC (Figure 15B) there was no change in the total protein content. The nonsignificant alteration was also observed in the rats treated with a combination of cartap aand malathion, when compared to the total protein content of rats treated with the individual pesticide (cartap and malathion). The percent decrease in the total protein content was computed to be 30.02, 37.05 and 24.74% in serum, 4.92, 6.01 and 9.84% in the membrane of PRBC, 28.85, 37.15 and 32.06% in brain (sucrose homogenate), 29.64, 33.03 and 30.27% in brain (homogenate prepared with the treatment of Triton X-100) and 21.69, 20.31 and 30.32% in liver tissue of their respective pesticide treated groups viz. cartap, malathion and cartap + malathion treated, respectively. When the pesticides treated rats were pretreated with the aqueous extract of *A. vera* leaves, these groups indicated complete protection from the pesticides induced alteration on the level of total protein content (Figure 3). The treatment of aqueous extract of *A. vera* leaves showed almost no change in this indice even up to 15 days.

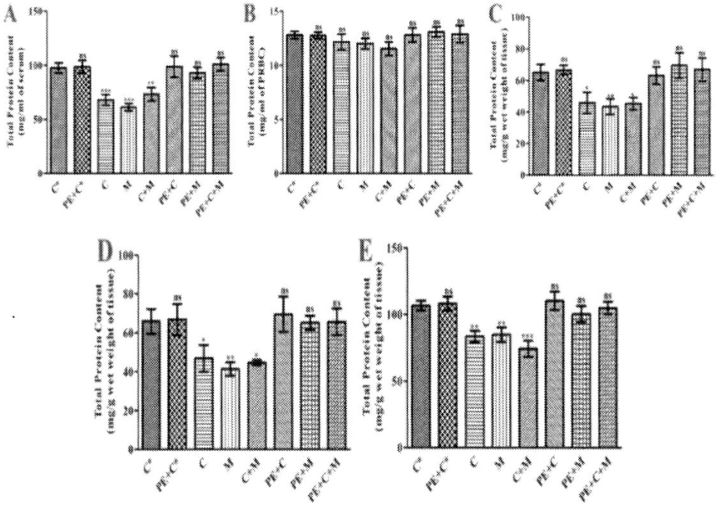

Figure 16. Effects of cartap, malathion and the aqueous extract of *A. vera* leaves on the level of protein content in (A) serum, (B) PRBC, (C) Brain (sucrose homogenate), (D) Brain (homogenate prepared with the treatment of Triton X-100) and (E) Liver. The procedures for the administration of pesticides and the aqueous extract of *A. vera* leaves as well as the estimation of prorein in serum and PRBC membrane were the same as described in Materials and Methods section. The unit was expressed as mg/ml in blood and mg/g wet weight of the tissue in liver and brain tissues. The data represented as mean ±SD of 3 independent experiments. The p values were non-significant at $p>0.05$ as compared to control group.

There are certain pesticides which have been reported to induce the DNA damage. The previous findings reported that the rats treated with cartap, malathion and their combination have strogly affected the oran and body weight of Wistar rats. To ascertain this, we endevoured to evaluate the effects of cartap, malathion and a combination of both pesticides on the levels of total DNA content in rat blood, brain and liver of Wistar rats. In order to evaluate the effect of these pesticides and the aqueous extract of *A. vera* leaves on the level of total DNA content, the procedures for the administration of pesticides, the preadministration of aqueous extract of *A. vera* leaves and the estimation of DNA content were the same as mentioned in Materials and Methods section. The results from present study indicated that the total DNA content in blood, brain and liver (Table 10) was nonsignificantly altered in the cartap and malathion treated rats. The nonsignificant alteration was also observed in the rats treated with a combination of cartap aand malathion, when compared to the total DNA content of rats treated with the individual pesticide (cartap and malathion). The treatment of aqueous extract of *A. vera* leaves showed almost no change in this indice even up to 15 days.

Table 10. Effects of cartap, malathion and aqueous extract *A. vera* leaves on the levels of total DNA content in rat blood, brain and liver in the different treatment groups

Experimental Groups	Blood (µg/ml)	Brain (µg/mg)	Liver (µg/mg)
C*	234.0±18.53	279.8±22.76	292.5±7.927
PE+C*	240.7±19.34 (+2.48%)	273.3±29.78 (-2.43%)	293.1±11.33 (-0.05%)
C	237.3±15.45 (+1.32%)	270.1±18.58 (-3.57%)	281.6±15.08 (-3.97%)
M	218.3±16.84 (-6.78%)	282.2±6.96 (+0.74%)	282.3±7.204 (-3.73%)
C+M	245.3±9.464 (+4.74%)	271.3±15.82 (-3.14%)	281.2±22.59 (-4.11%)
PE+C	255.1±14.34 (+8.92%)	279.4±17.02 (-0.25%)	274.6±14.07 (-6.36%)
PE+M	235.3±8.360 (+0.47%)	279.7±8.118 (-0.14%)	279.8±8.291 (-4.58%)
PE+C+M	231.6±8.493 (-1.10%)	272.8±8.476 (-2.61%)	277.4±12.91 (-5.40%)

(+) and (-) sign shows percent increase and decrease in the DNA content. The data represent mean ± SD of three independent experiments.

The pesticides have been reported to alter the ratio of protein and DNA in the mammalian systems. To ascertain this weendevoured to evaluate the effects of cartap, malathion and a combination of both pesticides on the ratio of total protein and DNA content in rat blood, brain and liver of Wistar rats. In order to evaluate the effect of these pesticides and the aqueous extract of *A. vera* leaves on the level of total DNA content, the procedures for the administration of pesticides, the preadministration of aqueous extract of *A. vera* leaves and the calculationof the ratio of protein and DNA content were the same as mentioned in Materials and Methods section. The results from present study indicated that the ratio of total protein and DNA content in blood, brain and liver was significantly decrased in the rats treated with cartap, malathion and a combination of cartap and malathion, when compared to the ratio of total protein and DNA content of control rats. The significant alteration was also observed in the rats treated with the combination of these pesticides when compared to the individual pesticides (cartap and malathion). The percent decrease in the ratio of total protein and DNA content was computed to be 33.73, 35.19 and 31.06% in blood, 19.69, 21.76 and 28.81% in brain, and 26.58, 35.45 and 29.97% in liver tissue of their respective pesticide treated groups viz. cartap, malathion and cartap+malathion treated, respectively. When the pesticides treated rats were pretreated with the aqueous extract of *A. vera* leaves, these groups indicated complete protection from the pesticides induced alteration on the level of total protein and DNA content ratio in blood, brain and liver (Table 11). The treatment of aqueous extract of *A. vera* leaves showed nonsignificant change in this indice even up to 15 days.

Table 11. Effects of cartap, malathion and aqueous extract *A. vera* leaves on the ratio of total protein and DNA content in rat blood, brain and liver in the different treatment groups

Experimental Groups	Blood (μg/ml)	Brain (μg/mg)	Liver (μg/mg)
C*	417.00	381.72	225.44
PE+C*	410.58 (-05.56%)	395.60 (-02.85%)	228.08 (-00.35%)
C	288.10 (-33.73%)*	308.88 (-19.69%)*	166.85 (-26.58%)*
M	281.74 (-35.19%)*	300.92 (-21.76%)*	146.70 (-35.45%)*
C+M	299.71 (-31.06%)*	273.80 (-28.81%)*	159.14 (-29.97%)*
PE+C	387.72 (-10.82%)	394.62 (+02.44%)	253.68 (+11.61%)
PE+M	396.17 (-08.88%)	359.49 (-17.31%)	233.58 (+02.77%)
PE+C+M	436.79 (-00.46%)	385.66 (-11.29%)	236.67 (+04.13%)

(+) and (-) sign shows percent increase and decrease, respectively, in the ratio of total protein and DNA content in rat blood, brain and liver in the different treatment groups. The data represent mean ± SD of three independent experiments. The p values significant (*) at $p < 0.05$ and non-significant (ns) at $p>0.05$ as compared to control group.

4.8. Effects of Pesticides and Aqueous Extract of *A. vera* Leaves on the Levels of Oxidative Stress Index (OSI) in Blood Serum, Brain and Liver of Wistar Rat

The pesticides have been reported to alter the ratio of pro-oxidant and antioxidant (i.e., OSI) in the mammalian systems including humans. To ascertain this, we endevoured to assess the effects of cartap, malathion and a combination of both pesticides on the ratio of pro-oxidant and antioxidant in rat blood, brain and liver of Wistar rats. In order to evaluate the effect of these pesticides and the aqueous extract of *A. vera* leaves on the level of pro-oxidant and antioxidant ratio or OSI the procedures for the administration of pesticides, the preadministration of aqueous extract of *A. vera* leaves and the calculation of the OSI were the same as mentioned in Materials and Methods section. The results from present study indicated that the OSI in blood, brain and liver was significantly decrased in the rats treated with cartap, malathion and a combination of cartap and malathion, when compared to the control rats. The percent decrease in the OSI was computed to be 17.69, 21.39 and 15.59% in blood, 57.85, 76.42 and 57.17% in brain, and 73.36, 90.45 and 273.94% in liver tissue of their respective pesticide treated groups viz. cartap, malathion, and cartap + malathion treated, respectively. When the pesticides treated rats were pretreated with the aqueous extract of *A. vera* leaves, these groups indicated complete protection from the pesticides induced alteration on the level of OSI in blood, brain and liver (Table 12). The treatment of aqueous extract of *A. vera* leaves showed nonsignificant change in this indice even up to 15 days.

Table 12. Effects of cartap, malathion and aqueous extract *A. vera* leaves on the level of oixidative stress index in rat blood, brain and liver in the different treatment groups

Experimental Groups	Blood (μg/ml)	Brain (μg/mg)	Liver (μg/mg)
C*	6.703	16.807	0.199
PE+C*	7.595 (+10.02%)	16.734 (-00.43%)	0.214 (+007.53%)
C	5.517 (-17.69%)	07.083 (-57.85%)	0.345 (+073.36%)
M	5.269 (-21.39%)	03.962 (-76.42%)	0.379 (+090.45%)
C+M	5.658 (-15.59%)	07.197 (-57.17%)	0.445 (+273.94%)
PE+C	7.035 (+04.95%)	17.674 (+05.15%)	0.222 (+011.55%)
PE+M	7.022 (+04.75%)	17.415 (+03.61%)	0.204 (+002.51%)
PE+C+M	7.995 (+19.27%)	17.456 (+03.86%)	0.234 (+017.58%)

(+) and (-) sign shows percent increase and decrease, respectively, in the OSI in rat blood, brain and liver in the different treatment groups. The data represent mean ± SD of three independent experiments. The p values significant (*) at $p < 0.05$ and non-significant (ns) at $p>0.05$ as compared to control group.

Chapter 5

Discussion

5.1. Aqueous Extraction of *A. vera* Leaves and Chemical Characterization of Extract

The dry matter that remains after the moisture analysis is called total solids (Nielsen, 2017). For various reasons, a good determination of the moisture content is still an essential and important element for the analytical procedures. This determination of the total amount of water is of great benefit to product manufacturers since water is an inexpensive load. The leaves of *A. vera* containing about 96% water content as observed in the present study is in good agreement with that reported by other workers (Eshun and He, 2004; Ahmed and Hussain, 2013; Boudreau et al., 2013; Benzidia et al., 2019). High water content in *A. vera* reflect on to its richness in mucilage which allows the retention of water.

The chemical analysis of *A. vera* leaves exhibited the presence of flavonoids, terpinoids, phenols, flavanol and carbohydrates similar to the findings reported by Raphael (2012) and Kumar et al., (2016). In addition to these, the presence of other constituents such as saponins have also been documented in the leaves of *A. vera* depending on the extraction solvents and methods. The presence of saponins are positive with ethanol, methanol, ethyl acetate, petroleum ether, acetone and hexane extracts and negative with the aqueous extract (Tripathi et al., 2013; Dharajiya et al., 2017). It is noted that *A. vera* plant is rich in secondary metabolites which are distinguished by some health promoting properties such as anti-inflammatory, antimalarial, antimicrobial, cytotoxicity and pharmacological effects (Iqbal et al., 2018). Tannins have a biological and pharmacological activity which includes antioxidative, antibacterial, antiviral, cardioprotective, antitumor, anti-inflammatory and immune-modulatory (Kumari and Jain, 2012). As for Flavonoids, they exhibit many pharmacological activities such as antioxidants, anti-allergic, anti-inflammatory, antimicrobial and anticancer properties (Khanam et al., 2015).

5.2. Evaluation of Antioxidant Potential of Plant Extract

Plant extracts including that of *A. vera* are known to contain several chemical constituents as demonstrated in above section. Phenolic compounds are known to exhibit strong antioxidant potential due to the presence of hydroxyl groups (Ozgen et al., 2010). Additionally, they are found to be effective in scavenging free radicals as a result of their redox properties that allow them to act as reducing agents (Oyedemi et al., 2010). Flavonoids are hydroxylated phenolics and are potent water-soluble antioxidants which help in radical scavenging and prevention of oxidative cell damage. They have been reported to possess strong antioxidant activities (Hesham et al., 2002; Loots et al., 2007). The phytochemicals investigated, as shown in this study, have been reported to possess strong antioxidant activities due to their ability to adsorb or quench free radicals and decompose peroxides generated in the system (Adedapo et al., 2008). The phytochemical evaluation of aqueous extract of *A. vera* showed the presence of high content of polyphenolic compounds such as total phenolic, flavonoids and flavanol content. These results were strongly corroborated with the findings of other researchers (Ahmed and Hussain, 2004; Wintola and Afolayan, 2011; Benzidia et al., 2018).

DPPH is a relatively stable free radical scavenger which converts the unpaired electrons to paired ones. The strong scavenging of DPPH radical by the aqueous extract of *A. vera* in this study indicates the potency of plant in donating hydrogen to the lone pair electron of the free radicals. Because the inhibition was more at a higher concentration in the extract, it could be suggested that the plant extract contain compounds capable of donating protons to the free radicals. This method has proven the effectiveness of the extract in a concentration-dependent manner. The DPPH assay of *A. vera* extract showed very high potential of free radical quenching activity which indicated the antioxidant potential of the extract with increasing order of the plant extract concentration. Our findings are similar to the findings of the other researchers (Wintola and Afolayan, 2011; Benzidia et al., 2018).

5.3. Morphological and Behavioural Symptoms Induced by Pesticides in Rats and Effect of Aqueous Extract of *A. vera*

Pesticides are considered as a double-edged weapon. Their application on one hand controls the insects/pests and helps in increasing the crop yields, whereas on the other hand they contaminate soil, water, air and the food stuffs. The indiscriminate use of pesticides has been resulted in several of toxicological consequences to the non target animals including the human health by affecting their key organs such as blood, brain, liver and kidney. The liver is the largest organ known for biotransformation of xenobiotics (Nelson et al., 1996). The exposure of living systems to the pesticides has been shown to impair the liver function and to cause injury to the hepatic tissues. Therefore, prevention and recovery of hepatic injury is a key to treat liver disease (El-Beshbishy, 2005). The application of plant extracts as therapeutic supplements has been reported to show great promises in this context (Madrigal-Santillan et al., 2014).

The body, organ and relative organ weight have been the most important morphological observation criteria for the determination of xenobiotics toxicity. The exposure of pesticides has been shown to cause morphological alterations in the mammalian systems (Saafi et al. 2011). In the present study the increase in the body weight of the Wistar rats treated with cartap, malathion and their combination were observed. The results of the present investigation have indicated significant alterations in the weights of the body and the organs i.e., brain and liver of rat treated with cartap and malathion, which could be due to occurrence of cholinergic deficiency and generation of oxidative stress as also reported by Saafi et al. (2011) and Gupta et al. (2019). Similar results have been shown by several workers, who exposed the rats with different classes of pesticides for varying treatment durations (Gupta et al., 2019). The present study also registered the alterations in the behavioural patterns such as hudling and cudling with mild tremor in the Wistar rats after pesticide exposure. Our findings are strongly corroborated by the previous results presented by other workers (Saafi et al. 2011; Gupta et al. 2019). However, the pre-treatment of animals with aqueous extract of *A. vera* leaves in the present study had potentialy attenuated the alterations of body and organ weights in pesticides treated rats, which may be due its high antioxidant effect of the plant extract.

5.4. Effects of Pesticides and Aqueous Extract of *A. vera* Leaves on the Status of Hematological Parameters of Rat Blood

The pesticide mediated alterations in different hematological parameters have been earlier indicated (Celik et al., 2009; Gupta et al., 2019). Our findings clearly showed that the pesticides' exposure reflected a non-significant change in these parameters excepting WBCs which significantly increased in pesticides treated rats (Figure 4). Similar to these results, Kalender et al., (2006) have also demonstrated increase in WBCs count in rats exposed with malathion. Though the rise in WBC count in the pesticides treated rats in present study was significant, the leucocytosis has been implicated to the severity of stress (Celik et al., 2009; Gupta et al., 2019). Thus, leukocytosis may reflect the response of immune system to the pesticide stress. In addition, OPs have been shown to induce the alteration in hematological parameter (Kalender et al., 2006; Elhalwagy et al., 2008; Gupta et al., 2019).

5.5. Effects of Pesticides and Aqueous Extract of *A. vera* Leaves on the Levels of Some Biochemical Paraameters in Rat Blood Serum

The exposure of animals to any xenobiotics results into generation of excess free radicals (ROS) which consequences into sharp decrease in the cellular antioxidants thereby inducing disordered redox system. The lipid peroxidation (LPO) is known to produce many thiobarbituric acid reactive substances (TBARS); the biomarkers of oxidative damage including MDA (Vincent and Taylor, 2006). The results generated from this study illustrate significantly enhanced MDA in rat serum when the animals were exposed to the pesticides. These findings were corroborated with findings of several workers (Hazarika et al., 2003; Selmi et al., 2018). However, this study indicated that when rats were pretreated with the aqueous extract of *A. vera* leaves, it significantly ameliorated rats from the pesticide induced toxicity as the MDA level came close to control. The aqueous extract of *A. vera* leaves have been shown to attenuate pesticides toxicity in rat serum due to their strong antioxidant potential (Werawatganon et al., 2014).

The GSH (tripeptide), is a potential cellular non-enzymatic biomarker of oxidative stress in the living systems (Ross, 1988). Our results showing significant reduction in the GSH content as recorded in the rat serum suggested oxidative stress development in the pesticide treated animals (Table 5). This finding is in agreement with the data reported by Selmi et al., (2018). The preadministration of aqueous extract of *A. vera* leaves, however, caused significant restoration of the GSH content in rat's blood serum treated with pesticides.

The serum levels of creatinine and uric acid is metabolic consequences of protein metabolism. The change in creatinine and uric acid level has been considered as the potential marker for the altered functions of kidney in mammals including rats and humans. These indices were assessed in current study allowed an indirect evaluation of renal function. The level of uric acid content in serum permits the assessment of basic protein metabolism through uric acid cycle (Gonzales et al., 1994). Once uric acid is evacuated principally in the blood by kidney, subsequently this approximately 40- 60% is reabsorbed at kidney tubules via glomerular filtration (Gonzales et al., 1994). In present investigation, blood uric acid was significantly elevated in rats treated with cartap, malathion and their mixture, suggesting renal damage. Another metabolite that was analyzed for the assessment of renal function in current study was creatinine. It is an excretory product of muscular activity which eliminated in blood exclusively by kidneys, therefore it has a strong association between the level of creatinine and kidney function. Mostly creatinine is removed by kidneys (filtered in renal glomeruli and tubules), is considered as the potential indicator for the renal-glomerular function (Benzidian et al., 2019). Similar to the level of uric acid, the level of creatinine expressively amplified in rat's blood serum that received cartap, malathion and their mixture, demonstrating the kidney damage in current study (Figure 6B). The level of uric acid and creatinine in blood serum have been used to detect renal injury. In the present study the significant increase in the level of uric acid and creatinine was observed which suggested that the pesticide perhaps caused nephrotoxicity in rats. Our finding is in strong corroboration with the earlier investigation which has shown increased levels of uric acid and creatinine in blood (Kanu et al., 2016). However, the pretreatment of aqueous extract of *A. vera* leaves expressively protected the uric acid and creatinine levels in rat exposed to cartap, malathion and their mixture. The renoprotective effect of *A. vera* might be useful in the treatment of cardiovascular diseases associated with kidney dysfunction along with the relevant therapeutics.

The enzymes of antioxidant defense system (GST, GPx, CAT and SOD) actively involved in maintaining the redox status and the detoxification of xenobiotics in the biological system. The glutathione-S-transferases (GSTs), catalyzes the reaction involving conjugation of reduced-glutathione to any xenobiotics leading to their detoxification (Padma et al., 2011) and the glutathione peroxidase (GPx) actively involved in the oxidation reduction reaction of GSH. The SOD, responsible to catalyze biochemical reaction involving dismutation of superoxide radicals to H_2O_2 which further degraded into O_2 and H_2O molecules by CAT (Halliwell, 1992). These enzymes are involved in instant removal of reactive oxygen species (ROS) produced due to the xenobiotic stress. Any perturbations in their activities caused by pesticide induced stress may result into the emergence of serious consequences at pathophysiological levels in mammalian system (Jaiswal et al., 2017; Gupta et al., 2019). Our results indicated notable changes in the activities of these enzymes due to pesticides exposure (Figure 3). These alterations in the functions of these enzymes have suggested the failure of antioxidant defense system. However, the pre-treatment of aqueous extract of *A. vera* leaves to rats reflected significant level of protection of these enzymes' activities. Our results are corroborated with the similar observations reported by other workers when rats were exposed with cartap, malathion and methicarb (Selmi et al., 2018; Gupta et al., 2019).

The levels of activities of transaminases (AST and ALT) and phosphatase (ALP) have been the potential biomarkers for assessing the liver damage/functions (Eraslan et al., 2009). Whenever the hepatocytes get damaged, several enzymes located inside including AST, ALT, ALP, and LDH are leaked out into the blood (Ncibi et al., 2008). The transaminases are known to play key roles in regulating physiological processes through catalyzing trans-amination reactions to facilitate metabolism of xenobiotics and other macromolecules (Mauro et al., 2006). LDH catalyzing the conversion of pyruvic acid into lactic acid in glycolysis has been used as a potential biomarker to assess the consequences of some pesticides (Eraslan et al., 2009; Jaiswal et al., 2016; Gupta et al., 2019). Jaiswal et al. (2016) have shown that the exposure of rats with carbofuran at their sublethal doses, another member of carbamate pesticide, can significantly alter the function of LDH in liver and heart of rats. In the present study a sharp increase in the activities of transaminases, phosphatases and LDH have been observed. The sharp enhancement in the activities of these enzymes indicated towards hepatic damage in rats treated with cartap, malathion and a combination of both pesticides. In another study the OPs have been demonstrated to induce the functions of LDH, ALP, ALT and AST (Celik et al., 2009). The reports from other workers have also indicated similar observations (Al-Shinnawy et al., 2014; Gupta et al., 2019). The results from our study have shown that the pretreatment with aqueous extract of *A. vera* leaves could significantly protect the activities of marker enzymes from the adverse impact of pesticides. The *A. vera* extract has been shown to contain a hepatoprotective impact against cartap and malathion induced toxicity in rats (Gupta et al., 2019). Similar results have also been reported by Al-Shinnawy et al. (2014) who have displayed the ameliorative potential of *A. vera* juice in young male rabbits.

The pesticides of CA and OP groups are considered to act as the notorious cholinergic inhibitors, due to their potential to inhibit AChE activity (Rai et al., 2009; Gupta et al., 2019). The hydrolysis of acetylcholine (ACh) by AChE is crucial for the nerve impulse transmission. The adverse effects of these pesticides in occupationally exposed people make it imperative to study the impact of both the pesticides using the specific activity of AChE as a surrogate biomarker of neurotoxicity in the membrane of erythrocytes isolated from Wistar rats. The findings from the present investigation indicated that AChE in erythrocytes of rat was membrane-bound and may be solubilized by Triton X-100. The membrane bound nature of AChE from RBC and brain tissues of rats has also been reported by Gupta et al., (2015) and Gupta et al., (2016), respectively. The results from present work indicated a significant reduction in the AChE activity from erythrocytes of rats exposed to both of the pesticides. However, the malathion was observed to inhibit the enzyme more strongly than the cartap because of covalent linkage of OPs with the hydroxyl group of serine residues present at the active site of enzyme (Selmi et al., 2018; Gupta et al., 2019). However, the rats pretreated with the aqueous extract of *A. vera* leaves could protect the activity of RBC's AChE from inhibitory effects of pesticides. No such reports have been made by other workers.

Proinflammatory cytokines are low molecular weight proteins secreted mainly by the activation of macrophages and lymphocytes (Benzer et al., 2018). Environmental contaminants, such as pesticides can induce the secretion of these cytokines (Gangemi et al., 2016; Mwanga et al., 2016). Dysregulation in the production of these cytokines has been associated with several pathological conditions such as inflammatory conditions and cancer (Multhoff et al., 2012). The xenobiotic induced oxidative stress is known to induce inflammatory responses via activation of redox-sensitive transcription factors, such as NF-κB (Forrester et al., 2018). In our study, these transcription factors could have been activated by the increased oxidative stress induced by cartap, malathion and a combination of both pesticides and observed the elevated levels of TNF-α and IL-6. The increased levels of TNF-α and IL-6 in pesticides-exposed rats thus indicates the generation of inflammation by cartap, malathion and acombination of both exposures. Furthermore, the present investigation revealed that ability of *A. vera* to attenuate cartap, malathion and a combination of both-induced inflammation; as the pre-treatment with the aqueous extract of *A. vera* leaves significantly reduced the levels of TNF- α and IL-6 in the pesticides-exposed rats.

The response against the inflammation occurs when the tissues are injured. Damaged cells release chemicals, including histamine, bradykinin, and prostaglandins, causing blood vessels to leak fluid into the tissues, thereby causing swelling. This helps to isolate the foreign substance from further contact with body tissues and enhance the phagocytosis. Cyclooxygenase (COX) enzymes (also known as prostaglandin-H2-synthases) act as catalysts in the production of prostaglandins (highly active pro-inflammatory mediators) from arachidonic acid during inflammatory processes. The inflammatory response may occur either by lipoxygenase pathway or by COX pathway. The glycoproteins from *A. vera* gel have been found to inhibit the activity of COX-2 (Yagi et al., 2003). The administration of *A. vera* has been shown to result in phagocytic and proliferative activity caused by the inhibition of COX pathways and reduction of prostaglandin

E2 (Vazquez et al., 1996; Lindsey et al., 2002,). The transcription levels of albumin and tumor necrosis factor (TNF)-genes were found to be involved in the early phase of acute inflammatory response. *A. vera* has been reported to contain the ability to inhibit the inflammatory process of burn injury (Duansak et al., 2009). The phytosterols (campesterol, β-sitosterol, lupeol, and cholesterol) present in *A. vera* has been found to be described as enhancer of the anti-inflamatory actions (Haller, 1990). The plant also contained anthraquinones and chromone, which possess strong anti-inflammatory effects on murine macrophages (Park et al., 2009). The extracts of *A. vera* prepared in ethyl acetate and ethanol were found to be helpful in the therapy of corneal inflammation (Wozzniak et al., 2012). Furthermore, in the process of inflammation, endothelium-derived mediators and complement factors stimulate leucocyte adhesion with the release of inflammatory mediators. Flavonoids are known to repress this leucocyte adhesion process thereby, preventing the release of oxidants and inflammatory mediators, and the consequent injury to tissues (Panche et al., 2016). The *A. vera* extract has been observed to contain the high content of flavonoids. Due to the high flavonoid content the aqueous extract of *A. vera* leaves may have the ability to suppress leucocyte adhesion and therefore, play additional role in its anti-inflammatory effect observed in this study. In addition, ingested aqueous extract of *A. vera* leaves may undergo chemical modifications and the resulting products contribute significantly to the biological effect along with the native phytochemicals.

5.6. Effects of Pesticides and Aqueous Extract of *A. vera* Leaves on the Levels of Some Biochemical Paraameters in the Brain of Wistar Rats

The wides pread use of pesticides has led to the exposure of animals, birds, and non-target invertebrates to these chemicals including the contamination of human food sources (Babu et al., 2006). Mammals are likely to be adversely affected by oral, dermal, and inhalation exposure to these pesticides (Lasram et al., 2009). The pesticides act as double-edged weapons i.e on the one hand it could control the target pests and help in increasing the crop yields, whereas on the other hand they contaminate soil, water, air and foodstuffs. The use of pesticides has been resulted in toxicological consequences on several non-target animals including the adverse impacts on human's health (Gupta and Sharma, 2016). The exposure of living systems to the pesticides has been shown to impair several functions of the organisms including the cholinergic, and energy metabolism in the rat brain. Therefore, prevention and recovery of these key enzymes activities needed to treat disease such as neurodegenerative diseases. The therapeutic application of plant extracts has been shown great promises in this context (Madrigal-Santillan et al., 2014).

The exposure to a xenobiotics or environmental pollutants have been reported to cause excess ROS generation, thereby establishing the oxidative stress which resulted into the decreased synthesis of antioxidants. The LPO produces several thiobarbituric acid reactive substances (TBARS) that are routinely monitored as the biomarkers of oxidative cellular damage including MDA (Vincent and Taylor, 2006). The results from present study clearly indicated that the exposure to sub-lethal doses of cartap, malathion and their combination significantly elevated the MDA level in the rat brain. The results from our study were corroborated with the findings of several other workers. The experimental rats exposed to cartap (Lakshmi et al., 2015) and malathion (Selmi et al., 2018) showed significant increase in the level MDA content in rat (Ahmed et al., 2000; John et al., 2001; Hazarika et al., 2003; Gutpta et al., 2019). However, the present investigation showed that the pretreatment of rats with aqueous extract of *A. vera* leaves was able to cause significant amelioration of rat brain from the pesticides toxicity as the level of MDA came close to normal. The aqueous extract of *A. vera* has also been reported to attenuate the oxidative toxicity in serum of rat's blood due to its strong antioxidant potential as reported elsewhere (Wintola and Afolayan, 2011; Werawatganon et al., 2014; Gupta et al., 2019).

The GSH, a tripeptide is a potential cellular non-enzymatic antioxidant biomarker of oxidative stress (Ross, 1988). In present investigation, the significant decrease in the level of GSH in rat brain reflected the establishment of oxidative stress in the cartap and malathion treated animals. The results from present study strongly corroborate with the results reported by several other workers (Ahmed et al., 2000; John et al., 2001; Hazarika et al., 2003; Selmi et al., 2018; Gupta et al., 2019). Similar observations have also been reported earlier in the brain of rat treated with different doses of carbofuran (Jaiswal et al., 2013). The pretreatment of aqueous extract of *A. vera* leaves significantly restored the GSH level in rat brain exposed to cartap and malathion as well as a mixture of both the pesticides.

The enzymes of glutathione redox system actively involved in maintaining the redox status as well as the detoxification of xenobiotics in the biological system. The Glutathione-S-transferases (GSTs), catalyzes the conjugation of reduced-GSH to xenobiotic substrates for their detoxification (Padma et al., 2011) and the glutathione peroxidase (GPx) is actively involved in the oxidation reduction reaction of GSH. These enzymes perform the instant removal/scavenging of the excess ROS produced during the oxidative process. In present investigation, the significant decrease in the activities of GST and increase in the GPx from rat brain exposed to the individual (cartap/malathion) and in combination of both pesticides, has suggested the failure of biological redox system due to the pesticide stress. The similar observations have been reported when rats were treated with other pesticides such as cartap (Gupta et al., 2019) malathion (Ahmed et al., 2000; Selmi et al., 2018; Gupta et al., 2019), methicarb (Ozden and Alpertunga, 2010) and carbofuran (Rai et al., 2009; Rai et al., 2011; Jaiswal et al., 2013). However, the pre-treatment of rats with the *A. vera* leaf extract restored the levels of these enzymes in the present study reflected significant level of amelioration from the pesticide toxicity.

The enzymes of the antioxidant defense system comprise SOD and CAT, which are responsible for neutralizing the free radicals into nontoxic chemical compounds. The SOD catalyzes the dismutation reaction of superoxide radicals into hydrogen peroxide which undergoes degradation reaction catalyzed by CAT, producing water and oxygen molecules (Halliwell, 1992). The alterations in the activity levels of these antioxidative enzymes due to the xenobiotics exposure may lead to the onset of serious consequences in the mammalian systems (Jaiswal et al., 2017). In present investigation, the activity levels of these enzymes were found to be significantlydecreased in rat brain due to the pesticide stress. The decreased activities of these enzymes from rat brain exposed to the individual and in combination of both the pesticides has indicated the failure of antioxidant system. Our results are in corroboration with the similar findings reported earlier (Ahmed et al., 2000; John et al., 2001; Gupta et al., 2019). However, the pre-treatment of rats with the *A. vera* leaf extract in the present investigation indicated the significant level of amelioration from the pesticide toxicity as the activities of SOD, and CAT appeared near to their control levels. No such reports have been made by other workers.

The LDH is a key cytosolic enzyme of energy metabolism which catalyzes the terminal step of glycolysis and has been implicated in the assessment of adverse impacts of pesticides on the energy metabolism of exposed animals (Jaiswal et al., 2016). Whenever the cell membrane is damaged, several enzymes located in the cytosol, including LDH, are leaked out into the blood (Ncibi et al., 2008). This enzyme is one of the potential biomarkers of cellular damage (Gokcimenet al., 2007; Eraslan et al., 2009) and the energy crisis. In the present study, the significant decrease in the LDH activity in rat's brain was observed, though no such studies have been made by other workers. The decrease in the LDH activity in rat brain as observed in the present study, may be due to the leaking of LDH into the blood (as the level of LDH was found to be increased in the blood serum). Though the exact mechanism of aaction of these pesticides on the LDH activity in rat is not known; it appears that both the pesticides may influences the enzyme activity via inhibition of pyruvate dehydrogenase complex (Jaiswal et al., 2016). However, the pretreatment of *A. vera* in the present investigation offered protection to the LDH activity from the pesticide toxicity.

The pesticides such as carbamates and organophosphates are the notorious inhibitors known to inhibit the activity of AChE (Rai et al., 2009). The hydrolysis of the neurotransmitter acetylcholine (ACh) to acetic acid and choline catalyzed by AChE is essential for the transmission of normal nerve impulses. The adverse effects of these pesticides make it imperative to study the impact of both the pesticides using the specific activity of AChE as a marker of neurotoxicity in the rat brain exposed to cartap, malathion and a combination of both the pesticides. The results of the present study indicated that AChE in rat brain was membrane-bound and could be solubilized by using Triton X-100. The similar reports have been reported, that AChE as membrane-bound form in the brain of rats (Gupta et al., 2016). In present investigation, the significant decrease in the activity of AChE from brain of rat exposed to the individual and in mixture of both the pesticides has suggested the strong inhibitory potential of these pesticides. However, the malathion was observed to inhibit the enzyme more potently than the cartap. The similar observations have been reported when rats were treated with other carbamate pesticides such as carbofuran (Rai et al., 2011; Jaiswal et al., 2013; Gupta et al., 2016) and malathion (John et al., 2001; Hazarika et al., 2003; Abdollahi et al., 2004; Selmi et al., 2018). However, the pretreatment of rats with the *A. vera* leaf extract in the present study reflected significant level of protection of AChE from the pesticide toxicity as the activities of this enzyme appeared close to the level of control in rat's brain under different treatment conditions. There were no reports available from other groups on the ameliorative impact of *A. vera* extract on the pesticide treated rat brain AChE.

5.7. The Pesticides Exposure Did Not Significantly Alter AChE Isozyme Functions *In Vivo*

Esterases are the enzyme proteins that are defined by their abilities to catalyze the hydrolysis of ester bonds in the compounds (El-Bermawy, 2004). Because most insecticides are esters of substituted phosphoric, carbamic, or cyclopropanecarboxylic acids, they are subjected to degradation by esterases (Devonshire, 1991). Therefore, esterases are the most significant enzymes for insecticides detoxification.

The results from present study revealed that the membrane bound AChE of the rat brain can be resolved into four isoenzymic forms using native-PAGE. Srinivasan et al. (1976) had demonstrated four isoenzymes in Triton

X-100 extract of rat brain synaptosomes and microsomes, while Vijayan and Brownson (1974) have reported three isoenzymes in Triton X-100 extract of the rat brain. Oderfeldnowak and Skangiel-Kramska (1976) and Rieger and Vigny (1976) have detected two AChE isoenzymes in the rat hippocampus after Triton X-100 treatment. By gel filtration, Wenthold et al. (1974) detected four different AChE forms. The comparison of these results is difficult because of different experimental conditions and enzyme sources. Detection of multiple molecular forms of AChE is well documented, but there is little information on the *in vivo* response of rat brain AChE isoenzymes against the pesticides.

In the present study, no significant change in the AChE isoenzyme activity to sublethal doses of cartap, malathion and a combination of both has been observed. Thus, no correlation between the levels of the control enzyme activity and the degree of isoenzyme inhibition could be demonstrated. There are only a few reports on AChE isoenzyme changes resulting from *in vivo* inhibition by organophosphates. Davis and Agranoff (1968) demonstrated the isoenzyme inhibition of rat retinal AChE by di-isopropyl phosphorofluoridate. The significance of isoenzyme behavior *in vivo* can only be speculated as very little information is available on the physicochemical basis and functional significance of rat brain AChE isoenzymes against these pesticides. It is possible that the sensitivity of isoenzymes may represent structural differences between molecules or groups of molecules governing the rates of reaction with these inhibitors. At the same time, it is also suggestive that such a response may be a function of specific spatial location of isoenzyme in relation to the nerve cell, with the most dominant and sensitive isoenzyme occupying sites easily accessible to the inhibitor. Previous reports on the AChE isozymes of rat brain have shown some discrepancies against parathion (Vijayan and Brownson, 1974).

Changes in the band intensity could be interpreted as a result of certain mutational events that would have occurred in the regulator genes, which would lead to inhibition, attenuation or constitutive gene expression. Therefore, the corresponding bands become faint or more intense. The recorded changes in band intensity could also be attributed to the cytological abnormalities induced by malathion treatments. Asita and Makhalemele

(2009) have concluded that the increase in band intensity could be due to gene duplication. Several experiments have shown that malathion affects activity of esterases (Flessel et al., 1993) suggesting that a metabolite of malathion (possibly the oxygen analog) was the actual inhibitor of malathion-esterase and that technical grade samples of malathion may contain a small quantity of this inhibitor. On the other hand, Shakoori and Saleem (1989) have revealed that there is general tendency towards enhanced enzyme activities after insecticide application, which could be perhaps due to increased concentration of enzymes following induction at the gene level.

5.8. Exposure of Pesticides May Cause Alterations in Protein Expression in Rat Brain; The Aqueous Extract of *A. vera* Leaves May Offer Amelioration

There are some studies which have indicated that the pesticides can decrease or change the levels of expression of different proteins in rat brain. Several enzymes which get inhibited by the pesticides, can modulate the function/expression of the genes. The results from the present study reflected perturbations in the expression of rat brain proteins due to pesticides exposure. In addition, some new protein bands had appeared as compared to the corresponding control. SDS protein profile showed 22 common bands in control, although there were changes in bands number and intensity. These results were in good agreement with the results of Zidan and Galal (2012), who reported the appearance of a definite decrease in protein bands number as a response to stress conditions. Whereas, there may be four basic differences in terms of protein content under the stress conditions: 1. production of some new proteins not present in the untreated group; 2. inhibition of some proteins that are produced by the untreated group; 3. increase in the level of expression of some proteins; 4. decrease in the level of expression of some proteins that are present in the untreated/control group. All the four differences were directly associated with the response of rat brain to the cartap and malathion treatments in the present investigation. The results were also in agreement with those of Rajendran et al. (2007), who reported that the production of novel proteins or the increased production of already existing proteins are due to stress response. In other words, the appearance of extra bands due to the treatment with an insecticide indicates that resulting proteins are probably responsible for the detoxification of the insecticide (Mohammed and Hafez, 2000). In addition, it was noticed that, no differences in the protein profile was observed in the rats treated with the combination of cartap and malathion, when compared with the individual treated pesticides. The alterations in the electrophoretic profiles of rat brain proteins are indicative of the ability of both cartap and malathion to alter the gene expression in treated animals. Among the two tested pesticides, cartap expressed the most deleterious effect on the protein synthesis, and remarkably inhibiting protein production in brain. Some electrophoretic bands could have disappeared due to the deletion of their corresponding genes (El-Khallal and Mohamed, 2004). On the other hand, appearance of new characteristic bands could be explained on the basis of mutational event at the regulatory system of unexpected gene(s) that activate it (El-Nahas, 2000). However, the disappearance of some bands in this study could be traced back due to the induction by aqueous extract of *A. vera* leaves. Further, the obtained results indicated that the stress protein analysis is a promising alternative and more sensitive method for measuring toxic effects on the organisms at sublethal levels. Thus it can be suggested that the proteomic profiling is a sensitive tool for environmental stress diagnosis and the stressed proteins could be used as biomarkers for environmental pollution identification. As reported before, the biomarkers of pesticides or specific pesticide exposure can be determined in part with gene expression profiling which has been suggested as a means of defining pesticide action and discovering pesticide alternatives (Duke et al. 2003).

5.9. Exposure to the Pesticides May Adversely Influence the Biochemical and Histological Parameters: Amelioration by Aqueous Extract of *A. vera* Leaves

The exposure to drugs, xenobiotics and environmental pollutants have been reported to cause generation of excess reactive oxygen species (ROS) thereby establishing the oxidative stress and decreased the synthesis of antioxidative agents. The oxidative process, such as lipid peroxidation (LPO) produces several compounds called as thiobarbituric acid reactive substances (TBARS) that are routinely used as biomarkers. MDA is one of them mostly used as indicators for the cellular redox status (Vincent and Taylor, 2006). The results from present study clearly indicated that the exposure to sub-lethal doses of cartap and malathion significantly elevated the level of MDA in rat liver showing increase in membrane lipid peroxidation. Our results were supported by the findings of other workers who exposed the rats to cartap (Lakshmi et al., 2015) and malathion (Moore et al., 2010; Arab et al., 2018; Selmi et al., 2018) and showed significant increase in MDA content in rat liver. However, in the present investigation, the pretreatment of rats with aqueous extract of *A. vera* leaves was able to cause significant amelioration of rat liver from the toxicity of pesticides as the level of MDA returned back to near normal. The aqueous extract of *A. vera* leaves has also been reported to attenuate the toxicity in liver of rats possibly due to its strong antioxidative potential as reported elsewhere (Wintola and Afolayan, 2011; Werawatganon et al., 2014).

The GSH, a tripeptide known as a potential cellular non-enzymatic antioxidant, has been recognized as another marker of oxidative stress in different tissues of living systems (Ross, 1988). Further, a significant inverse relationship between the extent of LPO production and GSH level in the rat liver has been demonstrated (Younes and Sieger, 1981). The significantly decreased level of GSH in rat liver as observed in the cartap and malathion treated rat reflected generation of oxidative stress in the animal. These findings were in agreement with the recent reports presented by some workers (Selmi et al., 2018; Arab et al., 2018). Similar observations have also been reported earlier in the brain of rat treated with different doses of carbofuran, another compound from the organocarbamate group of pesticides (Rai and Sharma, 2007; Jaiswal et al., 2013). The pretreatment of aqueous extract of *A. vera* leaves significantly restored the GSH level in rat liver exposed to cartap and malathion as well as a mixture of both. It was interesting to note that this ameliorative effect of plant extract was relatively less effective in the liver of rat treated with the malathion as compared to the cartap as well as the mixture of both of these pesticides.

The enzymatic antioxidant defense system in the living organisms comprise those enzymes which are responsible for neutralizing the free radicals or free radical producing molecules into nontoxic chemical species. These antioxidative enzymes are SOD, CAT and GST. SOD catalyzes the dismutation of superoxide radicals to hydrogen peroxide which undergoes degradation reaction catalyzed by CAT producing water and oxygen molecules (Halliwell, 1992). Glutathione-S-transferases (GSTs), previously known as ligandins, catalyzes the conjugation of reduced-GSH to xenobiotic substrates for their detoxification (Padma et al., 2011). Any alterations in the activity levels of these antioxidative enzymes due to the xenobiotics exposure may lead to onset of serious pathophysiological consequences in the tissues of key organs of mammalian system (Jaiswal et al., 2014; Jaiswal et al., 2017). In present investigation, the significant decrease in the activities of these antioxidative enzymes from rat liver exposed to the individual and in mixture of both the pesticides has suggested the failure of antioxidant defense system. The similar observations have been reported when rats were treated with other carbamate pesticides such as carbofuran (Rai et al., 2009; Rai et al., 2011; Jaiswal et al., 2013), methicarb (Ozden and Alpertunga, 2010) for longer duration and for an OP such as malathion (Selmi et al., 2018). In contrast, the activity of GST from the liver of rat exposed to chronic and acute doses of malathion have been demonstrated to be increased (Kaur and Sandhir, 2006). However, the pre-treatment of rats with the *A. vera* leaves extract in the present study reflected significant level of amelioration from the pesticide toxicity as the activities of SOD, CAT and GST were restored to near control levels.

The activities of transaminases (ALT and AST) have been considered as potential biomarkers of liver function test. While studying the impact of cartap, malathion and their combination on the rat liver function, the increased activities of these transaminases as observed in this investigation, were appeared as an indicative of damage to hepatic tissues. These findings are evidenced by other workers who have demonstrated perturbations in the activity levels of both these hepatic transaminases in rats exposed to carmate and organophosphate pesticides (Iseri et al., 2007; Liao et al., 2008; Al-Shinnawy et al., 2014). The *A. vera* extract have been shown to be a hepatoprotectant in terms of recovery in the activities of rat liver ALT and AST from malathion toxicity (Al-Shinnawy et al., 2014). The increased activities of transaminases in liver of mammalian systems have been shown to be associated with the input of amino acid into the tricarboxylic acid cycle (TCA) to overcome the energy crisis during pesticide stress (Phillp, 1995). The pretreatment of rats with aqueous extract of *A. vera* in present study was, however, found to modulate the effect of pesticides to bringing back the activities of ALT and AST to near normal levels.

The levels of ALP and ACP have been reported to be key markers for the assessment of liver functions. The activities of both phosphatases from the liver (Mauro et al., 2006) play critical role in regulation of biological processes such as the detoxification of xenobiotics, cellular metabolism and biosynthesis of energetic macromolecules. The significant increase in the activity of ALP and decrease in ACP activity from the liver of pesticide treated rat in present investigation indicated a strong impact of the exposure. Alterations in the activities of ALP and ACP has been associated with the damage or necrosis of hepatic tissues. These results are in consonance with the findings of other workers (Iseri et al., 2007; Liao et al., 2008; Al-Shinnawy et al., 2014). The malathion induced alterations in the activities of ALP and ACP has been reported to be ameliorated by *A. vera* leaves (Al-Shinnawy et al., 2014). The pretreatment of rats with aqueous extract of *A. vera* leaves helped maintain the activity level of alkaline phosphatase in liver of rat under different treatment conditions.

LDH catalyzes the terminal step of glycolysis and has been implicated in the assessment of adverse impacts of some pesticides on the energy metabolism of exposed animals (Jaiswal et al., 2016). The significant decrease in the LDH activity of rat's hepatic tissues as observed in the present study may be due to the indirect effect of pesticides as LDH is not the direct target of their actions. Jaiswal et al., (2013 and 2016) have reported similar results in the liver and heart of rats treated with carbofuran, another member of carbamate group of pesticides. However, the cartap caused increase in LDH activity as has been reported in cancer cell line (Liao et al., 2006). The administration of *A. vera* leaves has been reported to attenuate the alterations caused by malathion in rat serum (Al-Shinnawy et al., 2014). Though the exact action mechanism of these pesticides on the LDH activity in rat liver is not known, it appears that both the pesticides may influences the enzyme activity via inhibition of pyruvate dehydrogenase complex as reported previously (Jaiswal et al., 2016). However, the pretreatment of *A. vera* leaves in the present investigation offered complete protection of rat liver from the pesticide mediated toxicity as the activity of LDH was found to be near control. There is no report available from other workers in this context.

In the light of the above observations reflecting the impact of pesticides in rat liver, it was envisaged to study the status of rat liver tissues under these treatment conditions. The results from the present study indicated drastic alterations in the liver tissues due to pesticide exposure suggesting hepatotoxic effects of cartap, malathion and their combination. These findings received support from the results reported by Iseri et al. (2007). By conducting histological evaluation, they have shown that the exposure of rats to carbamate and organophosphate compounds caused acute activation of Kupffer cells, and degenerated hepatocytes. The protection from malathion induced histopathological alterations in rat liver have also been indicated by *A. vera* leaves (Selmi et al. 2018). Handy et al. (2002) have reported that the rats treated with malathion developed serious injuries into the liver cells. In the present study the protection of rat liver from pesticides intoxication was observed due to pretreatment of aqueous extract of *A. vera* leaves, which could be due to the presence of antioxidant property in herbal preparation of *A. vera* (Wintola and Afolayan, 2011).

5.10. Pesticides Treatment May Cause Perturbations in the Molecular Indices (Protein, DNA) in Rat: Amelioration by Aqueous Extracts of *A. vera* Leaves

The level of protein content in organisms is known to respond to a wide variety of stresses. The results from the present study indicated that the total protein was significantly decreased in the key organs (blood, brain and liver) of rats treated with cartap, malathion and a combination of both insecticides compared with the control. There was significant decrease in the malathion treated group in the blood, brain and liver. The decrease in the protein contents in rat tissues is an indicative to different pesticides exposure. The malathion treatment had induced a significant decrease in the protein content of the liver, testis and ovary as elucidated by Hassanein (1991). Our results were in agreement with those of Hussain et al. (2009), who reported that spinosad, an insecticide, decreased the total protein contents in adult beetles. The pesticides are known to inhibit the LDH which leads to the energy crisis in the cell. Thus, the reduction in the total protein content may be due to its break down to amino acids and entering them to TCA cycle as a keto-acid for energy production. The decreased level of protein content was significantly ameliorated by the preadministration of aqueous extract of *A. vera* leaves to the rats treated with pesticides. The protective mechanism of *A. vera* was not clealy known. The protection offered by *A. vera* may be due to its high antioxidant capacity.

The DNA content has been found not to be altered in all the treatments as compared to control in the blood, brain and liver of rats exposed to cartap, malathion and a combination of both pesticides (present study). It suggested that the cartap and malathion did not show any change in DNA content in the blood, brain and liver of the rats. Thus, it can be concluded that the non-significant alteration in the DNA content upon exposure to pesticides may be due to the cellular resistance to the DNA damaging effect in blood, brain and liver. In contrast to our findings, Zidan and Galal (2012) have reported a significant increase in the nucleic acid content in the malathion treated male Albino rats. There was not any change in the total DNA content was observed in the animals treated with the aqueous extract of *A. vera* leaves.

The ratio of total protein content and the total content in the blood brain and liver of rats treated with the cartap malathion and a combination of both pesticides was found to be altered significantly when compared with the control. The aqueous extract of *A. vera* leaves was able to offer complete protection from the pesticide induced change in the level of protein DNA ratio; the exact mechanism of action is not yet known.

5.11. Exposure to Pesticides Caused Alterations in Oxidative Stress Index (OSI) in Rat Tissues: Mitigation by Aqueous Extract of *A. vera* Leaves

The relationship between lipid peroxidation and antioxidant enzymes activities has been considered as an oxidative stress index (OSI) in an animal exposed to a xenobiotic. It has been best described by the ratio of P/A as explained elsewhere (Jaiswal et al., 2017). The values of this ratio denote an index of net free radical scavenging ability (Ling et al., 2010). The decreased values of P/A ratio in the blood serum, brain and increased value of P/A ratio in the liver of rat treated with cartap, malathion and a mixture of both the pesticides indicated increase in the level of prooxidant species and generation of ROS molecules imposing oxidative stress in the tissues. The higher P/A value for malathion or a mixture of cartap and malathion as compared to cartap suggested relatively stronger oxidative potential of malathion. This trend showed that the free radical generation in the blood serum, brain and liver of rat depends on the persistent nature of the pesticide. The data indicated that liver may be the highly susceptible organ to oxidative stress due to occurrence of the large amount of metabolic transformation reactions of xenobiotic molecules, the poor availability of antioxidants, and consumption of large amount of oxygen per unit weight (Lopez et al., 2017;

Gupta et al., 2019). The restoration of P/A ratio to near normal level upon the pretreatment of aqueous extract of *A. vera* leaves in the experimental animals treated with the pesticide suggested that this plant extract possessed potential antioxidative capability (Hu et al., 2003; Lopez et al., 2017).

Chapter 6

SUMMARY AND CONCLUSION

Pesticides may be defined as chemicals that are able to control insects and pests including weeds. These chemicals are widely used in the agricultural and health practices. The residual amounts of carbamate and organophosphate pesticides have been detected in the soil, water, vegetables, grains and other foods products. It has been reported that the exposure to the pesticide may induce toxicity in the exposed subjects which may manifest in the form of numerous adverse effects on health of the non-target living systems. The rise in pesticide applications has raised several environmental concerns. According to an estimate, only 2-5% of the total pesticides sprayed reach to the target and the rest reaches to different biotic (animals and plants) and abiotic (water, air and soil) components of the environment through the pesticide drift. These classes of pesticide have been shown to exert its toxic effects in various animal systems including the nervous system. Both of these groups of pesticides have a similar mode of action, affecting the nervous system by disrupting AChE activity. They disrupt the sodium/potassium balance of nerve fiber, following the forced them continuous conduction of nerve impulse which leads in developing several disease symptoms including weakness or paralysis. In spite of these the above-mentioned pesticides have also been reported to induce oxidative stress, and perturbations in energy metabolism of mammalian systems.

The chemicals enhancing the effect of other chemicals are called as synergists. The pesticides in combinations have been shown to induce synergistic effect by competing for a common target site. The exposure of mixture of pesticides is common in the aquatic environment, including lakes, river, streams and other surface water that support aquatic life. Assessing the cumulative toxicity of pesticides in combination has therefore been a challenge for environmental health assessment. The assumption of dose addition or concentration addition for mixtures of anticholinesterase pesticides has also been demonstrated previously by several workers. But the synergistic effect of cartap and malathion has not been attempted so far.

Oxidative stress reflects an imbalance between the systemic manifestation of reactive oxygen or reactive nitrogen species (ROS/RNS) and the biological system's ability to detoxify the reactive intermediates or to repair the resulting damage. Disturbances in the normal redox state of cells can cause toxic effects through the production of peroxides and free radicals that damage all components of the cell, including proteins, lipids, and DNA. Oxidative stress was shown to be induced by anthropogenic contaminants as persistent organic pollutants, heavy metals, and also by toxins produced during massive blooms of cyanobacteria. The protective systems like oxidized glutathione/reduced glutathione, glutathione reductase, catalase, superoxide dismutase and peroxidase activities, ascorbate and α-tocopherol, are all recognized as potential antioxidants and functions as a marker. Different xenobiotics can cause toxicity by acting on to different specific targets and lead to alterations in key biochemical parameters, known as biomarkers. However, the information regarding the specificity of biomarkers induced due to exposure to particular xenobiotics is still lacking. Biomarkers are the molecules that can be identified and monitored in different parts of the body. It refers to a broad subcategory of clinical signs. Biomarkers serve as indicators of a particular sensitivity of individuals to the effect of a xenobiotic or to the effects of a group of such compounds. Any variation from the normal level of a particular marker is indicative of abnormal function of the specific organ concerned which probably may lead to the onset of a disease. The biomarkers serve as clinical and diagnostic tools and as surrogate endpoints in clinical research. The increasing attention has been given towards the biomarkers of pesticides toxicity in mammalian systems. These markers are used to detect the adverse clinical impact of pesticides. These biomarkers are the measurable biochemical, physiological and the behavioral indices of animals such as inhibition of cholinesterase activity, inhibition of neuropathy target esterase activity, β-glucuronidase, cytogenetic biomarkers (aberrations in chromosomal, micronuclei, sister chromatid exchange, single cell gel electrophoresis or comet assay), delta-aminolaevulinic acid dehydratase, hematological parameters and non-enzymatic biochemical parameters such as MDA and GSH. Some enzymatic antioxidative parameters include superoxide dismutase, catalase, glutathione-S-transferase, glutathione peroxidase, and glutathione reductase. The phosphatases, transferases, and cytochrome P450 have also been considered as effective markers of pesticide stress.

The plant-based chemicals are also getting attention to be used as effective antioxidants. To alleviate the toxic impacts of various xenobiotics including the pesticides, the molecules derived from plants have shown great promises. The herbal medicines due to their low cost and least side effects have emerged as the richest source of antioxidants which have been exploited as viable supplements with the numerous therapeutics against various diseases including neurotoxicity, hepatotoxicity and cardiotoxicity. Most of the herbal products and/or herbal medicines including *A. vera* are the rich sources of antioxidants. *A. vera* with wide geographical distribution has been reported to contain large amount of various secondary metabolites and several natural antioxidant agents.

However, there has been lack of information regarding the impact of cartap, malathion and a combination of both the pesticides as well as the ameliorating effect of aqueous extract of *A. vera* leaves on some key biomarkers (biochemical and histological indices) in the blood, brain and liver of rats *in vivo*. In the light of above-mentioned knowledge gap, an endeavoure has been made to assess the impact of both the pesticides (individually as well as in combination) on the levels of oxidative stress mediated toxicity and its amelioration using aqueous extract of *A. vera* leaves.

The results from the present study have been summarized as following:

The young leaves of *A vera* (L.) plant (family Asphodelaceae) contained 96.4% of water content. The shade dried and powderized leaves were extracted using Soxhlet apparatus in aqueous solvent which yielded 3.38 g i.e., 6.76%. The chemical properties of *A. vera* leaves extract showed the presence of flavonoids, terpinoids, phenols and carbohydrates. However, the *A. vera* extract did not show the presence of tannins, saponins, steroids, quinones and alkaloids.

Therefore, the phytochemical evaluation of aqueous extract of *A. vera* leaves carried out indicated the presence of polyphenolic compounds such as total phenolic (332.67±78.43 i.e., 33.26%), flavonoids (43.87±47.87 i.e., 4.38%) and flavanol content (13.98±13.09 i.e., 1.39%) of the total dry weight of the *A. vera* leaves. The free radical scavenging potential of aqueous extract of *A. vera* leaves at different concentrations (25, 50, 100, 250, 500 and 1000 μg/ml) as determined by the DPPH assay showed that the extract was exhibiting free radical quenching potential in concentration dependent manner i.e. the scavenging activity of DPPH radical was higher with theincreasing concentration of the extract. The aqueous extracts of *A. vera* caused drastic decrease in absorbance of DPPH radical showing very high scavenging potential of the plant extract.

The morphological and behavioural alterations in rats exposed to sublethal concentration of cartap and malathion have been observed continuosly till 15 days. At the end of the treatment of pesticides upto 15 days, no death in any of the groups was observed. However, the pesticides treated groups showed appearance of varying degrees of morphometric and behavioural alterations. The symptoms included huddling and cuddling with mild tremor. When compared to control, the rat body weight due to pesticide treatment was significantly low. The average relative liver weight and average relative brain weight were significantly reduced in the rats treated with cartap, malathion and their combination. However, the pre-administration of *A. vera* extract to the pesticides treated groups registered an ameliorating impact on the changes in the morphological and behavioural indices.

Rats treated with the pesticides were also pre-administered orally with aqueous extract of *A. vera* leaves for the assessment of their protective effect on the hematological parameters, if any. The results demonstrated that the pesticide-treated rats did not show any significantly alterations in terms of any of the hematological parameters in comparison to control. The *A. vera* extract treatment also did not show any significant change in the hematological parameters among the rats from all the experimental groups, excepting the total WBC count which increased more than two-fold in the experimental groups of rats treated with malathion and a combination of cartap and malathion.

The effect of cartap, malathion and their combination, the male Wistar rats were exposed to these pesticides and *A. vera* extract. The results reflected that the treatment of rats with both of these pesticides caused significant ($P < 0.05$) increase in the MDA and decrease the GSH levels in rat serum brain and liver, when administered either individually or in combination. The aqueous extract of *A. vera* leaves did not exert any significant alterations in the level of MDA (3.73%) and GSH (5.30%). Each of these pesticides either used in individual (cartap/malathion) or in combination caused significant rise in the levels of MDA (i.e., 27.48, 42.52, and 25.33%, respectively) and decrease in the levels of GSH (i.e., 22.44, 26.12, 23.67%, respectively). The individual treatment with malathion exerted significantly higher level of change with increased MDA content (42.52%) and decreased GSH content (26.12%). When the rats were pretreated with the aqueous extract of *A. vera* leaves, their exposure to either cartap or malathion did not alter the levels of MDA and GSH as compared to control.

The pesticide treated rats were also registered significant ($p < 0.05$) increase in the serum level of creatinine and uric acid. Both pesticides used individually (cartap and malathion) or in combination caused noteworthy rise in the level of creatinine by 62.65, 79.57 and 65.71%, and uric acid by 19.29, 30.17 and 17.46% in their respective pesticide treated groups viz. cartap, malathion, and cartap+malathion, respectively. Malathion alone recorded comparatively higher consequence with the increased uric acid (30.17%) and creatinine (79.57%) level. The mixture of both of these pesticides did not show any significant change in the impact when compared to the effect of individual pesticides. The treatment of aqueous extract of *A. vera* leaves did not show any significant variation in the levels of creatinine (0.65%) and uric acid (1.82%). When the pesticides treated groups of animals were pretreated with the aqueous extract of *A. vera* leaves, the values of creatinine and uric acid were found to be almost near control.

Xenobiotics exposure is known to cause excessive production of free radicals and generation of oxidative stress. The treatment of rats with these pesticides induced significant elevation in the activities of GST, GPx, SOD and CAT in rat blood serum. The percent elevation in the activities of these enzymes was computed to be 73.82, 100.31 and 61.38% for GST; 65.19, 81.97, and 50.25% for GPx; 138.89, 216.80 and 115.58% for SOD, as well as 29.02, 43.36, and 28.54% for CAT in their respective pesticide treated groups viz. cartap treated, malathion treated and cartap + malathion treated, respectively. The mixture of both of these pesticides did not show any significant change in the impact when compared to the effect of individual pesticide. When rats were pretreated with the *A. vera* leaves extract, these groups indicated complete protection from the pesticides induced toxicity. The activities of GST, GPx, SOD and CAT were completely protected due to pretreatment of plant extract.

The phosphatases (ALP), transaminases (AST and ALT), and LDH are the cytosolic enzymes and considered as potential biomarkers of the status of hepatic function. The effect of sublethal doses of pesticides caused evaluation in the activities of ALP, ALT, AST and LDH in rat blood serum. The significant rise observed in the activities was as following: for AST by 92.10, 101.75 and 97.36%; ALT by 127.61, 153.33 and 131.11% and ALP by 70.07, 86.97 and 66.90%, respectively, in the cartap, malathion and cartap+malathion treated groups. The mixture of both of these pesticides did not show any improvement in the impact when compared to the effect of individual pesticide. The treatment of rats with these pesticides either individually or in combination resulted into significant reduction in the activity of LDH by 42.20, 58.54 and 40.26%, respectively, in the cartap, malathion and cartap+malathion treated groups as compared to that of the control. When compared to the impact of individual pesticides (cartap, malathion), the effect of their combination did not show any significant reduction in LDH activity. However, the pretreatment of aqueous extract of *A. vera* leaves indicated protective effect on rats from the pesticides induced toxicity.

The effect of sublethal doses of pesticides was evaluated by assaying the activity of AChE isolated from the rat blood erythrocyte's membrane indicated that the pesticides significantly ($p< 0.001$) inhibited the AChE activity by 52.84, 61.61 and 68.15%, respectively, in the cartap, malathion and their combination. When compared to the impact of individual pesticide, the effect of their combination did not show any more reduction in enzyme activity. The activity of erythrocyte's AChE was observed to be protected by the pre-administration of *A. vera* extract.

The serum level of proinflamatory cellular cytokines (TNF-α and IL-6) are considered as the potential markers for inflammation and cellular damage. The results from present study pro-inflammatory cytokines levels were increased significantly ($p < 0.05$) in the rats exposed to cartap, malathion and their combination. The concentrations of TNF-α and IL-6 in the serum were increased by 85.44, 163.76, and 80.77% for TNF-α as well as 53.86, 116.15, and 83.37% for IL-6, respectively in the experimental groups treated with cartap, malathion and their combination. When compared to the impact of individual pesticide, the effect of their combination did not show any more elevation in the levels of proinflamatory markers. The levels of these proinflamatory markers were observed to be protected by the pre-administration of *A. vera* extract.

In the above experiments we observed that biochemical changes in the rat erythrocyte membrane due to exposure of sublethal concentrations of cartap and malathion, which suggests that the pesticides through blood circulation system would also be reaching to the brain of rats and causing perturbations in its oxidative status. In order to evaluate the impact of cartap, malathion and a combination of both of these pesticides at their sublethal doses, the Wistar rats were treated for 15 days at the intervals of 24 h as described in Materials and Methods. The treated rats were also preadministered orally with aqueous extract of *A. vera* leaves to assess its protective effect, if any.

The results from this study indicated significant ($p < 0.05$) increase in the levels of MDA and decrease in the level of GSH in the rat brain when compared to that of the control. The pesticides (cartap and malathion) have caused significant increase in the level of MDA by i.e., 81.22, 102.88, and 80.29% and marked decrease in the level of GSH by i.e., 40.92, 57.08, and 48.52%, respectively, in the cartap, malathion and their combination in brain. The effect of malathion individually registered relatively higher level of alterations with the increased levels of MDA (102.88%) and decreased levels of GSH (57.08%). When compared to the impact of individual pesticides (cartap and malathion), the effect of their combination did not show any significant change in the levels of MDA and GSH. When the pesticide treated animals were subjected to the pretreatment of the aqueous extract of *A. vera* leaves, the values of MDA and GSH was found to be almost the same as that of the control.

The effects of pesticides on the activities of antioxidant enzymes viz; GST, GPx, SOD and CAT, in the homogenates of rat brain treated with cartap, malathion and a combination of both suggested the treatment of rats with these pesticides showed significant decrease in the activities of GST, SOD, and CAT except the activity of GPx which was observed to be increased signicicantly as compared to the control. The percent alteration in the activities of these enzymes was computed to be 30.53, 59.96 and 29.87% for GST, 35.62, 46.84 and 39.27% for GPx, 25.94, 35.26 and 22.16% for SOD, and 26.29, 39.07 and 27.90% for CAT in their respective pesticide treated groups viz. cartap treated, malathion treated and cartap+malathion treated, respectively. The mixture of both of these pesticides did not show any further change in the impact when compared to the effect of individual pesticide. When rats were pretreated with the *A. vera* leaves extract, these groups indicated complete protection from the pesticides induced toxicity. The activities of GST, GPx, SOD and CAT were completely protected due to pretreatment of plant extract.

The treatment of rats with these pesticides reflected significant decrease in the activity of rat brain LDH as compared to the control. The percent decrease in the activity of this enzyme was computed to be 36.16, 55.35 and 39.71%, respectively, in their respective pesticide treated groups viz. cartap, malathion, and cartap + malathion treated rats. The mixture of both of these pesticides did not show any significant change in the impact when compared to the effect of individual pesticide. When rats were pretreated with the *A. vera* leaves extract, these groups indicated complete protection from the pesticides induced toxicity. The activity of LDH was completely protected due to the pretreatment of plant extract.

The effects of sublethal doses of pesticides (individually and in combination of cartap and malathion) on the level of rat brain AChE activity indicated that the pesticides significantly ($p < 0.001$) inhibited the AChE activity by 29.42, 38.96 and 36.54%, respectively, in the cartap, malathion and their combination. When compared to the impact of individual pesticide, the effect of their combination did not show marked reduction in the enzyme activity. The activity of AChE extracted from rat brain was observed to be protected by the pre-administration of aqueous extract of *A. vera* leaves.

The status of rat brain AChE isozyme activity profile was also resolved in 7.5% native-PAGE gel. The PAGE analysis of AChE preparation showed a smear of bands suggesting the presence of four isozymes in rat brain treated with the cartap, malathion and a combination of both the pesticides. However, there was not any significant change observed in the isozyme activity of AChE extracted from the rat brain treated with the cartap, malathion and their combination.

The impact of cartap, malathion and a combination of both pesticides on the level of rat brain protein profile at their sublethal doses was determined. The results indicated that pesticide treatment caused variations in the intensity as well as the content of the protein bands in the cartap, malathion and a combination of cartap and malathion treated groups, respectively, as compared to the control. The combination of these two pesticides also did not showed any further changes in the protein profile when compared to the individual pesticide treated groups. The pretreatment of aqueous extract of *A. vera* leaves alone did not show any significant alteration in the level of protein profile in rat brain even after 15 days of treatment. Comparing the treatments with its corresponding control, several expressed protein bands were visualized after exposure to the cartap, malation and a combination of both in the rat's brain. However, the protein of molecular weight of 140 kDa was missing in all the treatment groups and 15 kDa protein in the cartap treated groups. The intensity of the 15 kDa protein was observed tobe decreased in malathion and a combination of cartap and malathion. In addition to several up and down regulated proteins, new protein bands were appeared when compared to the corresponding control. SDS protein profile showed 22 common bands in control and all treatments, although there were changes in bands intensity. Also, it could be observed that the treatment with these pesticides at sublethal concentration decreased bands number, this was constant with the results of total protein content which revealed the decrease compared with the control group.

The pesticides treated rats indicated significant ($p < 0.05$) rise in MDA and decrease in GSH in rat liver tissues when compared to that of control. The pesticides (cartap and malathion) have caused significant increase in the level of MDA by 25.19, 27.96 and 63.53%, and decrease in GSH level by 35.47, 43.13 and 39.40%, respectively, in the cartap, malathion and their combination. The effect of combination of these pesticides registered relatively higher level of alterations with the increased values of MDA (63.53%) and decreased level of GSH (39.40%). When compared to the impact of individual pesticides (cartap and malathion), the effect of their combination did not show any further change in the levels of MDA and GSH. However, when the pesticide treated animals were subjected to the pretreatment of the aqueous extract of *A. vera* leaves, the values of MDA and GSH were found to be almost the same as that of the control.

The effects of pesticides on the activities of antioxidant enzymes viz. SOD, CAT and GST, the assays for monitoring the activities of these enzymes were carried out using the homogenates of rat liver treated with cartap, malathion and a combination of both for 15 days at the intervals of 24 h. The results suggested that the treatment of rats with these pesticides showed significant reduction in the activities of SOD, CAT and GST as compared to the control. The percent alteration in the activities of these enzymes was calculated to be 34.53, 39.24 and 46.01% for SOD, 59.00, 45.75 and 51.46% for CAT, and 25.49, 31.52 and 24.41% for GST in their respective pesticide treated groups viz. cartap, malathion and cartap + malathion treated, respectively. The mixture of both of these pesticides did not show any significant change in the impact when compared to the effect of individual pesticide. When rats were pretreated with the aqueous extract *A. vera* leaves, these groups displayed complete protection of rat liver from the toxic effect of pesticides. The activities of SOD, CAT and GST were completely protected due to pretreatment of aqueous extract *A. vera* leaves.

The transaminases (AST and ALT) and phosphatases (ACP and ALP) are considered as potential biomarkers of the status of hepatic function. The effect of sublethal doses of pesticides on the rat liver function was evaluated. The results indicated significant increase in the activities of AST by 38.57, 47.67 and 31.76% and ALT by 25.91, 35.10 and 24.08% and ALP by 31.92, 42.77 and 30.72% and significant decrease in the levels of ACP by 29.82, 32.10 and 28.94% in the rat liver, respectively, in the cartap, malathion and cartap+malathion treated groups. The mixture of both these pesticides did not show any significant change in the impact when compared to the effect of individual pesticide. When rats were pretreated with the aqueous extract *A. vera* leaves, these groups displayed complete protection of rat liver function from the hepatotoxic effect of the pesticides. The activities of AST, ALT, ACP and ALP were completely protected due to pretreatment of aqueous extract *A. vera* leaves.

The effects of pesticides on the activity of LDH was monitoring using the homogenates of liver tissue isolated from rats reated with cartap, malathion and a combination of both. The results suggested that the treatment of rats with these pesticides cased a significant reduction in the LDH activity as compared to the control. The percent decrease in the activity of this enzymes was computed to be 26.44, 33.59 and 33.00%, respectively, in their respective pesticide treated groups viz. cartap, malathion and cartap + malathion treated, respectively. The mixture of both of these pesticides did not show any significant change in the impact when compared to the effect of individual pesticide. When rats were pretreated with the *A. vera* leaves extract, these groups indicated complete protection from the pesticides induced toxicity. The activity of LDH was completely restored to the normal level due to pretreatment of plant extract.

To examine the effects of pesticides on the levels of histological parameters in the rat liver tissues were processed and analyzed. The microscopic examination of the tissue slides indicated the normal histological architecture of liver to contain acini and hexagonal lobules. The hexagonal lobules were centered on central vein and had portal triad containing hepatic artery, portal vein and bile duct. The data revealed the appearance of mild disorganization of the central vein and hepatic cords in the rat liver tissues treated with cartap, malathion, and their combination. Liver sections from the pretreatment of rats with aqueous extract of *A. vera* leaves exhibited moderate recovery/protection from the pesticide induced damage.

The previous findings reported that the rats treated with cartap, malathion and their combination have strogly affected the biochemical and morphological i.e. weight of oran and body, parameters of Wistar rats. To ascertain this, we endevoured to evaluate the effects of cartap, malathion and a combination of both pesticides on the levels of total protein content in rat blood (serum and PRBC), brain and liver of Wistar rats. The results from present study indicated that the total protein content in serum, brain and liver was significantly decrased whereas in case of PRBC there was no change in the total protein content was observed. The non-significant alteration was also observed in the rats treated with a combination of cartap aand malathion, when compared to the total protein content of rats treated with the individual pesticide (cartap and malathion). The percent decrease in the total protein content was computed to be 30.02, 37.05 and 24.74% in serum, 4.92, 6.01 and 9.84% in the membrane of PRBC, 28.85, 37.15 and 32.06% in brain (sucrose homogenate), 29.64, 33.03 and 30.27% in brain (homogenate prepared with the treatment of triton X-100) and 21.69, 20.31 and 30.32% in liver tissue of their respective pesticide treated groups viz. cartap, malathion, and cartap + malathion treated, respectively. When the pesticides treated rats were pretreated with the aqueous extract of *A. vera* leaves, these groups indicated complete protection from the pesticides induced alteration on the level of total protein content. The treatment of aqueous extract of *A. vera* leaves showed almost no change in this indice even up to 15 days.

There are certain pesticides which have been reported to induce the DNA damage. The previous findings reported that the rats treated with cartap, malathion and their combination have strogly affected the oran and body weight of Wistar rats. To ascertain this, we endevoured to evaluate the effects of cartap, malathion and a combination of both pesticides on the levels of total DNA content in rat blood, brain and liver of Wistar rats. The results from present study indicated that the total DNA content in blood, brain and liver was nonsignificantly altered in the cartap and malathion treated rats. The nonsignificant alteration was also observed in the rats treated with a combination of cartap and malathion, when compared to the total DNA content of rats treated with the individual pesticide (cartap and malathion).

The pesticides have been reported to alter the ratio of protein and DNA in the mammalian systems. The results from present study indicated that the ratio of total protein and DNA content in blood, brain and liver was significantly decrased in the rats treated with cartap, malathion and a combination of cartap and malathion, when compared to the ratio of total protein and DNA content of control rats. The significant alteration was also observed in the rats treated with the combination of these pesticides when compared to the individual pesticides (cartap and malathion). The percent decrease in the ratio of total protein and DNA content was computed to be 33.73, 35.19 and 31.06% in blood, 19.69, 21.76 and 28.81% in brain, and 26.58, 35.45 and 29.97% in liver tissue of their respective pesticide treated groups viz. cartap, malathion and cartap + malathion treated, respectively. When the pesticides treated rats were pretreated with the aqueous extract of *A. vera* leaves, these groups indicated complete protection from the pesticides induced alteration on the level of total protein and DNA content ratio in blood, brain and liver.

The pesticides have been reported to alter the ratio of pro-oxidant and antioxidant (i.e. OSI) in the mammalian systems including humans. To ascertain this, we endevoured to assess the effects of cartap, malathion and a combination of both pesticides on the ratio of pro-oxidant and antioxidant in rat blood, brain and liver of Wistar rats. The results from present study indicated that the OSI in blood, brain and liver was significantly decrased in the rats treated with cartap, malathion and a combination of cartap and malathion, when compared to the control rats. The percent decrease in the

OSI was computed to be 17.69, 21.39 and 15.59% in blood, 57.85, 76.42 and 57.17% in brain, and 73.36, 90.45 and 273.94% in liver tissue of their respective pesticide treated groups viz. cartap, malathion and cartap+malathion treated, respectively. When the pesticides treated rats were pretreated with the aqueous extract of *A. vera* leaves, these groups indicated complete protection from the pesticides induced alteration on the level of OSI in blood, brain and liver.

Taken together the results from the present study on the oral exposure of subacute doses of cartap and malathion at the intervals of 24 h upto 15 days of duration showed that these pesticides were able to establish the oxidative stress in rat blood, brain and liver. The results of the present study concerning the exposure of cartap, malathion and a combination of both pesticides at their sublethal doses indicated significant alterations on the levels of different biochemical, molecular and histological markers in the experimental rats, which could be ameliorated by the pre-administration of aqueous extract of *A. vera* leaves. The protecting impact of aqueous extract of *A. vera* leaves may be due to their strong antioxidant potential. The different phytochemicals present in *A. vera* could be employed as a therapeutic supplement along with proper medication against pesticide poisoning.

REFERENCES

Abdollahi M., Mostafalou S., Pournourmohammadi S., Shadnia S., (2004) Oxidative stress and cholinesterase inhibition in saliva and plasma of rats following subchronic exposure to malathion, *Comp Biochem Physiol C Toxicol Pharmacol,* 137:29-34.

Adams NH., Levi PE., Hodgson E., (1993) Regulation of cytochrome P-450 isozymes by methylenedioxyphenyl compounds, *Chem Biol Interact,* 86:255-274.

Adedapo AA., Jimoh FO., Afolayan AJ., Masika PJ., (2008) Antioxidant activities and phenolic contents of the methanol extracts of the stems of *Acokanthera oppositifolia* and *Adenia gummifera*, *BMC Complement Altern Med,* 8:54–60.

Afshar S., Farshid AA., Heidari R., Ilkhanipour M., (2008) Histopathological changes in the liver and kidney tissues of Wistar albino rat exposed to fenitrothion, *Toxicol Ind Health,* 24:581-586.

Agrawal A., Sharma B., (2010) Pesticides induced oxidative stress in mammalian systems, *Int J Biol Med Res,* 1:90-104.

Agrawal A., Sharma B., (2012) Natural products and their antioxidant potential, *Ind J Nat Prod Res,* 8:72-87.

Ahlawat KS., Khatkar BS., (2011) Processing, food applications and safety of *Aloe vera* products: A review, *J Food Sci Tech,* 48:525-533.

Ahmed M., Hussain F., (2013) Chemical composition and biochemical activity of *Aloe vera* (*Aloe barbadensis* Miller) leaves, *IJCBS* 3:29–33.

Ahmed RS., Seth V., Pasha ST., Banerjee BD., (2000) Influence of dietary ginger (*Zingiber officinales* Rosc) on oxidative stress induced by malathion in rats, *Food Chem Toxicol,* 38:443-450.

Aitken RJ., De Iuliis GN., Gibb Z., Baker MA., (2012) The Simmet lecture: new horizons on an old landscape-oxidative stress, DNA damage and apoptosis in the male germ line, *Reprod Domest Anim,* 47:7–14.

Alberts BJA., Julian L., David M., Martin R., Keith R., Peter W., (2014) *Molecular Biology of the Cell* (6th ed.), New York: Garland Science, 1025.

Al-Shinnawy MS., Hassan AR., Ismail DA., Shahin MA., (2014) The potential protective and therapeutic effects of *Aloe vera* juice on malathion induced hepatotoxicity in rabbits, *Egypt J Hosp Med,* 55:146-158.

Ames BN., (1983) Dietary carcinogens and anticarcinogens. Oxygen radicals and degenerative diseases, *Science,* 221:1256–64.

Arab SA., Nikravesh MR., Jalali M., Fazel A., (2018) Evaluation of oxidative stress indices after exposure to malathion and protective effects of ascorbic acid in ovarian tissue of adult female rats, *Electr phy,* 10:6789-6795.

Arner ES., Holmgren A., (2000) Physiological functions of thioredoxin and thioredoxin reductase, *Eur J Biochem,* 267:6102–9.

Asita AO., Makhalemele R., (2009), Genotoxic effects of dithane, malathion and garden ripcord on onion root tip cells, *AJFAND online,* 9:1191-1209.

Avijgan M., Mahboubi M., Moheb NM., Ahmadi NE., Yousefi H., (2014) Synergistic activity between *Echinophoraplatyloba* DC ethanolic extract and azole drugs against clinical isolates of *Candida albicans* from women suffering chronic recurrent vaginitis, *J Mycol Med,* 24:112-116.

Babu SN., Malik JK., Rao GS., (2006) Effects of subchronic malathion exposure on the pharmacokinetic disposition of pefloxacin, *Env Toxicol Pharma,* 22:167-171.

Badreddine A., (2016) *Préparation et Caractérisation d'extraits d'Argania Spinosa et d'huile d'argan et Evaluation de Leurs Effets Neuroprotecteurs In Vivo et In Vitro* [*Preparation and Characterization of Argania Spinosa and Argan Oil Extracts and Evaluation of Their Neuroprotective Effects In Vivo and In Vitro*]. Ph.D. Thesis, Université de Bourgogne, Dijon, France; Université Hassan ier, Settat, Morocco, p100.

Baghersad S., Bahrami SH., Mohammadi MR., Mojtahedi MRM., Milan PB., (2018) Development of biodegradable electrospun gelatin/aloe-vera/poly (ε caprolactone) hybrid nanofibrous scaffold for application as skin substitutes, *Mater Sci Eng C Mater Biol Appl,* 93:367-379.

Bannister JV., Bannister WH., Rotilio G., (1987) Aspects of the structure, function, and applications of superoxide dismutase, *CRC Crit Rev Biochem,* 22:111–80.

Barron KD. Bernsohn J., (1968) Esterases of developing human brain, *J Neurochem,* 15:273- 284.

Barth M., (1970) *Master's Thesis*, University of Michigan, Michigan.

Bartling A., Worek F., Szinicz L., Thiermann H., (2007) Enzyme-kinetic investigation of different sarin analogues reacting with human acetylcholinesterase and butyrylcholinesterase, *Toxicol,* 233:166-172.

Beers JrRF., Sizer IW., (1952) A spectrophotometric method for measuring the breakdown of hydrogen peroxide by catalase, *J Biol Chem,* 195:133-40.

Belikov AV., Schraven B., Simeoni L., (2015) T cells and reactive oxygen species, *J Biomed Sci,* 22: 85.

Benford DJ., Hanley AB., Bottrill K., Oehlschlager S., Balls M., Branca, F., Castegnaro JJ., Descotes J., Hemminiki K., Lindsay D., Schilter B., (2000) Biomarkers as predictive tools in toxicity testing, *Altern Lab Anim,* 28:19-131.

Benzer F, Kandemir FM, Kucukler S, Comaklı S, Caglayan C. Chemoprotective effects of curcumin on doxorubicin-induced nephrotoxicity in wistar rats: by modulating inflammatory cytokines, apoptosis, oxidative stress and oxidative DNA damage. *Arch Physiol Biochem.* 2018 Dec;124(5):448-457.

Benzidia B., Barbouchi M., Hammouch H., Belahbib N., Zouarhi M., Erramli H., Naima AD., Badrane N., Hajjaji N., (2019) Chemical composition and antioxidant activity of tannins extract from green rind of *Aloe vera* (L.) Burm. (F.), *J King Saud Univ-Sci,* 31(4):1175-1181.

Bernsohn J., Barron KD., (1964) Multiple molecular forms of brain hydrolases, *Int Rev Neurobiol,* 7, 297-344.

Bernstein H., Bernstein H., Payne CM., Dvorak K., Garewal H., (2008) Field defects in progression to gastrointestinal tract cancers. *Can Lett,* 260(1-2): 1-10.

Bonner MR., Coble J., Blair A., Beane Freeman LE., Hoppin JA., Sandler DP., Alavanja

Booth J., Boyland E., Sims P., (1961) An enzyme from rat liver catalysing conjugations with glutathione, *Biochem J,* 79:516–524.

Boudreau MD., Beland FA., (2006) An evaluation of the biological and toxicological properties of *Aloe barbadensis* (miller), *Aloe vera*, *J Environ Sci Health C Environ Carcinog Ecotoxicol Rev* 24:103-154.

Boudreau MD., Mellick PW., Olson GR., Felton RP., Thorn BT., Beland FA., (2013) Clear evidence of carcinogenic activity by a whole-leaf extract of *Aloe barbadensis* Miller (*Aloe vera*) in F344/N rats, *Toxicol Sci,* 131(1):26–39.

Boudreau MD., Mellick PW., Olson GR., Felton RP., Thorn BT., Beland FA., Clear evidence of carcinogenic activity by a whole-leaf extract of *Aloe barbadensis* miller (*Aloe vera*) in F344/N rats, *Toxicol Sci,* 131(1):26-39.

Brooker RJ., (2011) *Genetics: analysis and principles* (4th ed.), McGraw-Hill Science.

Burton K., (1956) A study of the conditions and mechanism of the diphenylamine reaction for the colorimetric estimation of deoxyribonucleic acid, *Biochem J,* 62:315-323.

Cairns RA., Harris IS., Mak TW., (2011) Regulation of cancer cell metabolism, *Nat Rev Cancer,* 11:85–95.

Cao BJ., Chen ZK., Chi ZQ., (1990) Antidotal effects of sulfhydryl compounds on acute poisonings by sodium ammonium dimethyl-2-

(propane-1,3-dithiosulfate) monohydrate, nereis toxin and cartap, *Zhongguo Yao Li Xue Bao*, 11:180-4.

Carney JM., Starke-Reed PE., Oliver CN., Landum RW., Cheng MS., Wu JF., Floyd RA., (1991) Reversal of age-related increase in brain protein oxidation, decrease in enzyme activity, and loss in temporal and spatial memory by chronic administration of the spin-trapping compound N-tert-butyl-alpha-phenylnitrone, *Proc Nat Acad Sci USA,* 88:3633–3636.

Carvalho CD., Costa HS., Albuquerque TG., Ramos F., Castilho MC., Sanches SA., (2015) Advances in phenolic compounds analysis of aromatic plants and their potential applications, *Trends Food Sci Tech,* 45:336-354.

Cedergreen N., (2014) Quantifying synergy: a systematic review of mixture toxicity studies within environmental toxicology, *PLoS One,* 9:1-12.

Cedergreen N., Dalhoff K., Gottardi, DLM., Kretschmann AC., (2017) Can toxicokinetic and toxicodynamic modeling be used to understand and predict synergistic interactions between chemicals? *Environ Sci Technol,* 51:14379-14389.

Celik I., Yilmaz Z., Turkoglu V., (2009) Hematotoxic and hepatotoxic effects of dichlorvos at sublethal dosages in rats, *Environ Toxicol,* 24:128-132.

Chandegara V., Varshney A., (2013) *Aloe vera* L. processing and products: a review, *Int J Aromatic Plants,* 3:492-506.

Chaudière J., Ferrari-Iliou R., (1999) Intracellular antioxidants: from chemical to biochemical mechanisms, *Food Chem Toxicol,* 37:949–62.

Chomczynski P, Sacchi N., (1987) Single step method of RNA isolation by acid guanidinium thiocyanate phenol-chloroform extraction, *Annal Biochem,* 162:156-159.

Chow JTN., Williamson DA., Yates KM., Goux WJ., (2005) Chemical characterization of the immunomodulating polysaccharide of *Aloe vera* L, *Carbohyd Res, 3*40:1131-1142.

Claiborne A., Yeh JI., Mallett TC., Luba J., Crane EJ., Charrier V., Parsonage D., (1999) Protein-sulfenic acids: diverse roles for an unlikely player in enzyme catalysis and redox regulation, *Biochem,* 38:15407-16.

Clause BT., (1998) *The Wistar Institute Archives: rats (not mice) and history, Mendel Newsletter*, Archived 2006 at the Wayback Machine.

Colovic MB., Krstic DZ., Lazarevic-Pasti TD., Bondzic AM., Vasic VM. (2013) Acetylcholinesterase inhibitors: pharmacology and toxicology, *Curr Neuropharmacol,* 11:315-35.

Conner GE., Salathe M., Forteza R., (2002) Lactoperoxidase and hydrogen peroxide metabolism in the airway, *Am J Respir Crit Care Med,* 166:57–61.

Curtin JF., Donovan M., Cotter, TG., (2002) Regulation and measurement of oxidative stress in apoptosis, *J Immunol Met,* 265:49–72.

Da Silva AP., Meotti FC., Santos AR., Farina M., (2006) Lactational exposure to malathion inhibits brain acetylcholinesterase in mice, *Neurotoxicol,* 27:1101-1105.

Dabelstein W., Reglitzky A., Schütze A., Reders K., (2007) Automotive Fuels. *Ullmann's Encyclopedia of Industrial Chemistry.*

Davidson A., (1999) *The Oxford companion to food.* Oxford, UK: Oxford University Press.

Davis GA., Agranoff BW., (1968) Metabolic behaviour of isozymes of acetylcholinesterase. *Nature,* 220: 277.

D'Cruz PM., Yasumura D., Weir J., Matthes MT., Abderrahim H., LaVail MM., Vollrath D., (2000) Mutation of the receptor tyrosine kinase gene Mertk in the retinal dystrophic RCS rat, *Human Mol Gene,* 9:645–51.

Deffert C., Cachat J., Krause KH., (2014) Phagocyte NADPH oxidase, chronic granulomatous disease and mycobacterial infections, *Cell Microbiol,* 16:1168–78.

del Río LA., López-Huertas E., (2016), ROS generation in peroxisomes and its role in cell signaling, *Plant Cell Physiol,* 57:1364–1376.

del Río LA., Sandalio LM., Palma JM., Bueno P., Corpas FJ., (1992) Metabolism of oxygen radicals in peroxisomes and cellular implications, *Free Rad Biol Med,* 13:557–80.

Devasagayam TP., Tilak JC., Boloor KK., Sane KS., Ghaskadbi SS., Lele RD., (2004) Free radicals and antioxidants in human health: current status and future prospects, *J Assoc Physicians India,* 52: 794–804.

Devonshire AL., (1991) Role of esterases in resistance of insects to insecticides, *Biochem Soc Trans,* 19:755-759.

Dharajiya D., Pagi N., Jasani H., Patel P., (2017) Antimicrobial Activity and Phytochemical Screening of *Aloe vera* (*Aloe barbadensis* Miller), *Int J Curr Microbiol Appl Sci* 6:2152–2162.

Dickinson BC, Chang CJ., (2011) Chemistry and biology of reactive oxygen species in signaling or stress responses, *Nat Chem Biol,* 7:504–11.

Dietz KJ., Jacob S., Oelze ML., Laxa M., Tognetti V., de Miranda SM., Baier M., Finkemeier I., (2006) The function of peroxiredoxins in plant organelle redox metabolism, *J Exp Bot,* 57:1697–709.

Dragland S., Senoo H., Wake K., Holte K., Blomhoff R., (2003) Several culinary and medicinal herbs are important sources of dietary antioxidants, *J Nutr,* 133:1286–1290.

Duansak D., Somboonwong J., Patumraj S., (2009) Effect of *Aloe vera* on leukocyte adhesion and TNF-a and IL-6 levels in burn wounded rats, *Med Biochem Biophys,* 29:239-246.

Duke SO., Baerson SR., Dayan FE., Rimando AM., Scheffler BE., Tellez MR., Wedge DE., Schrader KK., Akey DH., Arthur FH., De Lucca AJ., Gibson DM., Harrison HF., Peterson JK., Gealy DR., Tworkoski T., Wilson CL., Morris JB., (2003) United States Department of Agriculture-Agricultural Research Service, Research on natural products for pest management, *Pest Manag Sci,* 59: 708-717.

El-Bermawy SM., (2004) Esterase patterns in *Spodoptera littoralis* (Boisd.) after botanical extract treatments, *J Egypt Ger Soc Zool,* 43: 119-136.

El-Beshbishy HA., (2005) Hepatoprotective effect of green tea (*Camellia sinensis*) extract against tamoxifen-induced liver injury in rats, *J Biochem Mol Biol,* 38:563-570.

Elhalwagy MEA., Darwish NS., Zaher EM., (2008) Prophylactic effect of green tea polyphenols against liver and kidney injury induced by fenitrothion insecticide, *Pestic Biochem Physiol,* 9: 81-89.

El-Khallal SM., Mohamed THR, (2004) Changes in growth protein pattern, DNA finger prints and chromosomal aberrations of water stressed maize seedlings treated with abscisic or jasmonic acid, *Egypt J Biotechnol,* 18:32-338.

Ellman DL., Courtnedy K., Andres V., Featherstone RM., (1961) A new and rapid colorimetric determination of acetylcholinesterase activity. *Biochem. Pharmac.* 7:88-95.

Ellman GL., (1959) Tissue sulfhydryl groups, *Archives of Biochemistry and Biophysics,* 82:70-77.

Ellman GL., Courtney KD., Andres Jr V., Featherstone RM., (1961) A new and rapid colorimetric determination of acetylcholinesterase activity. *Biochemical pharmacol,* 7:88-95.

El-Nahas AI., (2000) Mutagentic potential of imazethapyr herbicide (pursuit) on *Vicia faba*. In the presence of urea fertilizers, *Pak J Biol Sci,* 3: 900-905.

Eranko Q., Kokkoa. & Soderholum. (1962a) *Ann. Med Exp. Bid. Fenn.* 40, 146-161.

Eraslan G., Kanbur M., Silici S., (2009) Effect of carbaryl on some biochemical changes in rats: the ameliorative effect of bee pollen, *Food Chem Toxicol,* 47:86-91.

Erink O., Kokko A., Soderholm U., (1962) Starch electrophoretic separation of esterases in the adrenal medulla, adrenal cortex and the brain of the rat and the hamster. *Ann Acad Sci Fenn,* 40: 146-161.

Ernst E., (2000) Adverse effects of herbal drugs in dermatology, *Brit J Dermatol,* 143:923-929.

Eshun K., He Q., (2004) *Aloe vera*: a valuable ingredient for the food, pharmaceutical and cosmetic industries a review, *Crit Rev Food Sci Nutr,* 44:91-96.

Estabrook RW (2003) A passion for P450s (remembrances of the early history of research on cytochrome P450), *Drug Metab Dispos,* 31:1461–1473.

Estabrook, RW., Cooper, DY., Rosenthal O., (1963) The light reversible carbon monoxide inhibition of steroid C-21 hydroxylase system in adrenal cortex. *Biochem Z,* 338:741-755.

Femenia A., Sanchez ES., Simal S., Rossello C., (1999) Compositional features of polysaccharides from *Aloe vera* (*Aloe barbadensis* Miller) plant tissues, *Carboh Pol* 39:109-117.

Ferreira GK., Carvalho-Silva M., Goncalves CL., Vieira JS., Scaini G., Ghedim FV., Deroza PF., Zugno AI., Pereira TCB., Oliveira GMT., Kist LW., Bogo MR., Schuck PF., Ferreira GC., Streckab EL., (2012) L-tyrosine administration increases acetylcholinesterase activity in rats. *Neurochem Int,* 61:370-1374.

Festing MF., May D., Connors TA., Lovell D., Sparrow S., (1978) An athymic nude mutation in the rat, *Nature.* 274:365–6.

Flessel P., Quintana PE., Hooper K., (1993), Genetic toxicity of malathion: a review, *Environ Mol Mutagen,* 22: 7-17.

Forrester SJ., Kikuchi DS., Hernandes MS., Xu Q., Griendling KK., (2018) Reactive oxygen species in metabolic and inflammatory signaling, *Circul Res,* 122:877–902.

Franco JL., Posser T., Mattos JJ., (2009) Zinc reverses malathion-induced impairment in antioxidant defenses, *Toxicol Lett,* 187:137-143.

Fukuto TR., (1990) Mechanism of action of organophosphorus and carbamate pesticides, *Env Health Persp,* 87:245-54.

Fulda S., Galluzzi L., Kroemer G., (2010) Targeting mitochondria for cancer therapy, *Nat Rev Drug Discov*, 9:447–64.

Gangemi S., Gofita E., Costa C., Teodoro M., Briguglio G., Nikitovic D., Tzanakakis G., Tsatsakis AM., Wilks MF., Spandidos DA., Fenga C., (2016) Occupational and environmental exposure to pesticides and cytokine pathways in chronic diseases (review). *Int J Mol Med,* 38(4), 1012-1020.

Gangemi S., Gofita E., Costa C., Teodoro M., Briguglio G., Nikitovic D., Tzanakakis G., Tsatsakis AM., Wilks MF., Spandidos DA., Fenga C., (2016) Occupational and environmental exposure to pesticides and cytokine pathways in chronic diseases (Review). *Int J Mole Med,* 38(4), 1012–1020.

Ghosh AK., Banerjee M., Mandal TK., Mishra A., Bhowmik MK., (2011) A study on analgesic efficacy and adverse effects of *A. vera* in Wistar rats, *Pharmacolonline,* 1:1098-1108.

Gibson SB., (2010) A matter of balance between life and death: targeting reactive oxygen species (ROS)-induced autophagy for cancer therapy, *Autophagy,* 6:835–7.

Gilden RC., Huffling K., Sattler B., (2010) Pesticides and health risks, *J Obstet Gynecol Neonatal Nurs,* 39:103–10.

Gilliom RJ., (2007) Pesticides in U.S. streams and ground water, *Environ Sci Technol,* 41:3407-3413.

Gokalp O., Buyukvanli B., Cicek E., (2005) The effects of diazinon on pancreatic damage and ameliorating role of vitamin E and vitamin C, *Pestic Biochem Phys,* 81:123-128.

Gokcimen A., Gulle K., Demirin H., Bayram D., Kocak A., Altuntas I., (2007) Effects of diazinon at different doses on rat liver and pancreas tissues, *Pestic Biochem Physiol,* 87:103-108.

Goldberg AF., Barka T., (1962) Acid phosphatase activity in human blood cells, *Nature,* 195, 297.

Goldman LR., (2004) *Childhood Pesticide Poisoning: Information for Advocacy and Action*, WHO.

Goldman LR., (2007) Managing pesticide chronic health risks: U.S. policies, *J Agromedicine,* 12:67–75.

Goodman B., (2011) Pesticide Exposure in Womb Linked to Lower IQ. *Health & Pregnancy.*

Gupta RC., Goad JT., Kadel WL., (1994) In vivo acute effects of carbofuran on protein lipid and lipoproteins in rat liver and serum, *J Toxicol Environ Health,* 42:451-62.

Gupta SC., Hevia D., Patchva S., Park B., Koh W., Aggarwal BB., (2012) Upsides and downsides of reactive oxygen species for cancer: the roles of reactive oxygen species in tumorigenesis, prevention, and therapy, *Antiox Red Signal,* 16:295–322.

Gupta VK., Pal R., Siddiqi NJ., Sharma B., (2015) Acetylcholinesterase from human erythrocytes as a surrogate biomarker of lead induced neurotoxicity, *Enz Res,* 7.

Gupta VK., Pathak A., Siddiqi NJ., Sharma B., (2016) Carbofuran modulating functions of acetylcholinesterase from rat brain in vitro, *Adv Biol,* 3760967, 7.

Gupta VK., Sharma B., (2015) Environmental hazards due to xenobiotics contamination: growing risk to human health and possible remedies,

Green Social Work: Environmental Protection, Petagon Press, ISBN 978-81-8274-852-1.

Gupta VK., Sharma B., (2016) Alleviation of acute poisoning of organophosphates in humans, *Anatomy Physiol Biochem Int J,* 1:555558.

Gupta VK., Sharma B., (2017) Role of phytochemicals in neurotrophins mediated regulation of Alzheimer's disease, *Int J Complement Altern Med,* 7:00231.

Gupta VK., Siddiqi NJ., Ojha AK., Sharma B., (2019) Hepatoprotective effect of *Aloe vera* against cartap- and malathion-induced toxicity in Wistar rats, *J Cell Physiol,* 234:18329-18343.

Gupta VK., Singh S., Agrawal A., Siddiqi NJ., Sharma B. (2015) Phytochemicals mediated remediation of neurotoxicity induced by heavy metals., *Biochem Res Int,* 9.

Guz J., Gackowski D., Foksinski M., Rozalski R., Zarakowska E., Siomek A., Szpila A., Kotzbach M., Kotzbach R., Olinski R., (2013) Comparison of oxidative stress/DNA damage in semen and blood of fertile and infertile men, *PLoS One,* 8:e68490.

Habeeb F., Shakir E., Bradbury F., Cameron P., Taravati MR., Drummond AJ., Gray AI., Ferro VA., (2007) Screening methods used to determine the anti-microbial properties of *Aloe vera* inner gel, *Methods,* 42:315-320.

Habig WH., Pabst MJ. Jakoby WB., (1974) Glutathione S-transferases. The first enzymatic step in mercapturic acid formation, *J Biol Chem,* 249:7130-7139.

Haller JS., (1990) A drug for all seasons, medical and pharmacological history of aloe, *Bull NY Acad Med,* 66:647-659.

Halliwell B., (1992) Reactive oxygen species and the central nervous system, *J Neurochem,* 59:1609-1623.

Hamman JH., (2008) Composition and applications of *Aloe vera* leaf gel, *Mole,* 13:1599-1616.

Hampton MB., Orrenius S., (1997) Dual regulation of caspase activity by hydrogen peroxide: implications for apoptosis, *FEBS Lett,* 414:552–6.

Han D., Williams E., Cadenas E., (2001) Mitochondrial respiratory chain-dependent generation of superoxide anion and its release into the intermembrane space, *Biochem J,* 353:411–6.

Handy RD., Abd-El Samei HA., Bayomy MFF., (2002) Chronic diazinon exposure: pathologies of spleen, thymus, blood cells, and lymph nodes are modulated by dietary protein or lipid in the mouse, *Toxicol,* 172:13-34.

Hanukoglu I., Rapoport R., Weiner L., Sklan D., (1993) Electron leakage from the mitochondrial NADPH-adrenodoxin reductase-adrenodoxin-P450scc (cholesterol side chain cleavage) system, *Arch Biochem Biophys,* 305:489–98.

Hassanein HMA., (1991) *Biological studies on the effect of some water pollutants (pesticides) on freshwater fish, Gambusia affinis*, M. S. Thesis, Fac Sci (Sohag) Assiut Univ. Egypt.

Hayes JD., Flanagan JU., Jowsey IR., (2005) Glutathione transferases, *Ann Rev Pharmacol Toxicol,* 45:51–88.

Hayes, WJJr., and Laws ERJr., (Eds.) *Handbook of Pesticide Toxicology. Vol. 2. Classes of Pesticides*. Academic Press, Inc. San Diego, CA. 1527p.

Hayyan M., Hashim MA., Al-Nashef IM., (2016) Superoxide ion: generation and chemical implications, *Chem Rev,* 116: 3029–85.

Hazarika A., Sarkar SN., Hajare S., Kataria M., Malik JK., (2003) Influence of malathion pretreatment on the toxicity of anilofos in male rats: a biochemical interaction study, *Toxicol,* 185:1-8.

Heath DF., (1961) *Organophosphorous Poisons*. Pergamon, Oxford.

Hesham J., El-Seedi R., Vishiyama S., (2002) Chemistry of bioflavanoids, *Ind J Pharm Edu,* 36:191–4.

Hiner AN., Raven EL., Thorneley RN., García-Cánovas F., Rodríguez-López JN., (2002) Mechanisms of compound I formation in heme peroxidases, *J Inorg Biochem,* 91:27–34.

Ho YS., Magnenat JL., Bronson RT., Cao J., Gargano M., Sugawara M., Funk CD., (1997), Mice deficient in cellular glutathione peroxidase develop normally and show no increased sensitivity to hyperoxia, *J Biol Chem,* 272:16644-51.

Ho YS., Magnenat JL., Gargano M., Cao J., (1998) The nature of antioxidant defense mechanisms: a lesson from transgenic studies, *Environ Health Persp,* 106:1219–28.

Hodgson AV., White TB., White JW., Strobel HW., (1993) Expression analysis of the mixed function oxidase system in rat brain by the polymerase chain reaction. *Mol Cell Biochem,* 120:171-179.

Horecker BL. Kornberg A., (1948) The extinction coefficient of the reduced band of pyridine nucleotides, *J Biol chem,* 175: 385-90.

Hossain MS., Mamun-Or-Rashid ANM., Towfique NM., Sen MK., (2013) A review on ethnopharmacological potential of *Aloe vera*, L. *J Intercult Ethnopharm,* 2:113-120.

Hsu CC., Lai SC., (2007) Matrix metalloproteinase-2, -9 and -13 are involved in fibronectin degradation of rat lung granulomatous fibrosis caused by *Angiostrongylus cantonensis*, *Int J Exp Pathol,* 88:437–43.

Hu Y., Xu, J., Hu Q., (2003) Evaluation of antioxidant potential of *Aloe vera* (*Aloe barbadensis* miller) extracts, *J Agric Food Chem,* 1:7788-91.

Hussain R., Ashfaq M., Saleem MA., (2009) Biochemical abnormalities produced by spinosad in *Tribolium castaneum* adult beetles, *Int J Agric Biol,* 11:241-244.

Hutter JA., Salman M., Stavinoha WB., Satsangi N., Williams RF., Streeper RT., Weintraub ST., (1996) Anti-inflammatory C-glucosyl chromone from *Aloe barbadensis*, *J Nat Prod,* 59:541-543.

International Code of Conduct on the Distribution and Use of Pesticides, *UN FAO*, 2008.

International Committee on Standardized Genetic Nomenclature for Mice/Rat Genome and Nomenclature Committee (2016), *Rules and Guidelines for Nomenclature of Mouse and Rat Strains, Mouse Genome Informatics*. Jackson Laboratory.

Iqbal E., Lim LBL., Salim KA., Faizi S., Ahmed A., Mohamed AJ., (2018) Isolation and characterization of aristolactam alkaloids from the stem bark of *Goniothalamus velutinus* (Airy Shaw) and their biological activities, *J King Saud Univ Sci,* 30:41–48.

Iqbal Z., Talwagr P., (1971) Acetylcholinesterase in developing chick embryo brain, *J Neurochem,* 18, 1261-1267.

Irani K., Xia Y., Zweier JL., Sollott SJ., Der CJ., Fearon ER., Sundaresan M., Finkel T., Goldschmidt-Clermont PJ., (1997) Mitogenic signalling mediated by oxidants in Ras-transformed fibroblasts, *Science,* 275:1649–52.

Iseri S., Ercan F., Gedik N., Yuksel M., Alican I., (2007) Simvastatin attenuates cisplatin-induced kidney and liver damage in rats, *Toxicol,* 230:256-264.

Jaiswal SK., Gupta VK., Ansari MD., Siddiqi NJ., Sharma, B., (2017) Vitamin C acts as a hepatoprotectant in carbofuran treated rat liver slices in vitro, *Toxicol Rep,* 4:265-273.

Jaiswal SK., Gupta VK., Siddiqi NJ., Sharma B., (2017) Curcumin mediated attenuation of carbofuran induced toxicity in the heart of Wistar rats, *Cell Mol Biol* (Noisy-le-grand), 63:12-17.

Jaiswal SK., Sharma A., Gupta VK., Singh RK., Sharma B., (2016) Curcumin mediated attenuation of carbofuran induced oxidative stress in rat brain, *Biochem Res Int,* 7637931.

Jaiswal SK., Siddiqi NJ., Sharma B., (2013) Carbofuran imbalances of the redox status in rat brain: amelioration by vitamin C, *J Biochem Res,* 1:26-35.

Jakoby WB., Ziegler DM., (1990) The enzymes of detoxication, *J Biol Chem,* 265:20715–20718.

Jani GK., Shah DP., Jain VC., Patel MJ., Vithalan DA., (2007) Evaluating mucilage from *Aloe Barbadensis* Miller as a pharmaceutical excipient for sustained-release matrix tablets, *Pharma Tech,* 31:90-98.

Jeyaratnam J., (1990). Acute pesticide poisoning: a major global health problem, World Health Statistics Quarterly. *Rapport Trimestriel de Statistiques Sanitaires Mondiales,* 43:139–44.

Jinesh GG., Taoka R., Zhang Q., Gorantla S., Kamat AM., (2016) Novel PKC-ζ to p47 phox interaction is necessary for transformation from blebbishields, *Sci Rep,* 6:23965.

John S., Kale M., Rathore N., Bhatnagar D., (2001) Protective effect of vitamin E in dimethoate and malathion induced oxidative stress in rat erythrocytes, *J Nutri Biochem,* 12:500-504.

Johnson F., Giulivi C., (2005) Superoxide dismutases and their impact upon human health, *Mol Asp Med,* 26:340–52.

Johri A., Dhawan A., Singh RL., Parmar D., (2006) Effects of prenatal exposure of deltamethrin on the ontogeny of xenobiotic metabolizing cytochrome P450s in the brain and liver of offsprings. *Toxicol Appl Pharmacol.,* 214:279-289.

Johri A., Yadav S., Dhawan A., Parmar D., (2007) Over expression of cytochrome P450s alters behavioral activity of rat offspring following prenatal exposure to lindane. *Toxicol Appl Pharmacol,* 225:278-292.

Jones WD., Sacamano C., (2000) *Landscape plants for dry regions: more than 600 species from around the world.* Calif Bill's Autom Pub, USA. 38-40.

Jonsson TJ., Lowther WT., (2007) The peroxiredoxin repair proteins, *Subcell Biochem,* 44:115-41.

Joseph B., Raj SJ., (2010) Pharmacognostic and phytochemical properties of *Aloe vera* Linn- an overview, *Int J Pharma Sci Rev Res,* 4:106-110.

Jurewicz J., Hanke W., (2008) Prenatal and childhood exposure to pesticides and neurobehavioral development: review of epidemiological studies, *Int J Occup Med Environ Health,* 21:121-32.

Kalender Y., Uzunhisarcikli M., Ogutcu A., Acikgoz F., Kalender S., (2006) Effects of diazinon on pseudocholinesterase activity and haematological indices in rats: the protective role of vitamin E, *Environ Toxicol Pharmacol,* 22:46-51.

Kalender Y., Uzunhisarcikli M., Ogutcu A., Acikgoz F., Kalender S., (2006) Effects of diazinon on pseudocholinesterase activity and haematological indices in rats: the protective role of vitamin E, *Environ Toxicol Pharmacol,* 22:46-51.

Kamboj M., Mahajan S., (2007) Micronucleus-an upcoming marker of genotoxic damage, *Clinic oral investig,* 11:121–126.

Kamiya A., Inoue Y., Gonzalez FJ., (2003) Role of the hepatocyte nuclear factor 4α in control of the pregnane X receptor during fetal liver development, *Hepatol,* 37:1375-1384.

Kamrin MA., (1997) *Pesticide profiles: Toxicity, environmental impact, and fate* (1st ed.). Boca Raton: CRC. ISBN 978-1566701907.

Kar SK., Bera TK., (2018) Phytochemical constituents of *Aloe vera* and their multifunctional properties: A comprehensive review, *Int J Pharm Sci Res,* 9:1416-23.

Kaur M., Sandhir R., (2006) Comparative effects of acute and chronic carbofuran exposure on oxidative stress and drug metabolizing enzymes in liver, *Drug Chem Toxicol,* 29: 415-421.

Kava R., Greenwood MR., Johnson PR., (1990), Zucker (fa/fa) *Rat. ILAR Journal, Inst Lab Res,* 32:4–8.

Khan SM., Sobti RC., Kataria L., (2005) Pesticide-induced alteration in mice hepatooxidative status and protective effects of black tea extract, *Clinica Chimia Acta; Int J Clinic Chem,* 358:131–138.

Khanam Z., Wen CS., Bhat IU., (2015) Phytochemical screening and antimicrobial activity of root and stem extracts of wild *Eurycoma longifolia* Jack (Tongkat Ali), *J King Saud Univ Sci* 27:23–30.

Kidd H., James DR., (1987) *The Agrochemicals Handbook.* 2nd ed. Lechworth, Herts, England: The Royal Society of Chemistry, 1987, p. A614.

Kikkawa S., Yamamoto T., Misaki K., Ikeda Y., Okado H., Ogawa M., Woodhams PL., Terashima T., (2003) Missplicing resulting from a short deletion in the reel in gene causes reeler-like neuronal disorders in the mutant shaking rat Kawasaki, *J Comp Neurol,* 463:303–15.

Kim H., Panteleyev AA., Jahoda CA., Ishii Y., Christiano AM., (2004) Genomic organization and analysis of the hairless gene in four hypotrichotic rat strains, *Mamm Genome,* 15:975–81.

Kim HJ., Kim CH., Ryu JH., Kim MJ., Park CY., Lee JM., Holtzman MJ., Yoon JH., (2013) Reactive oxygen species induce antiviral innate immune response through IFN-λ regulation in human nasal epithelial cells, *Am J Respir Cell Mol Biol,* 49:855–65.

Kimura H, Suzuki A (eds.). *New Research on DNA Damages.* New York: Nova Science Publishers, Inc. pp. 1–47.

Kumar S., Budhwar L., Yadav A., Yadav M., Yadav JP., (2016) Phytochemical screening and antibacterial activity of *Aloe vera* collected from different climatic regions of India, *Nat Prod J* 6:73–82.

Kumaran A., Karunakaran RJ., (2007) In vitro antioxidant activities of methanol extracts of *Phyllanthus* species from India, *LWT- Food Sci Technol,* 40:344-352.

Kumari M., Jain S., (2012) Tannins: an antinutrient with positive effect to manage diabetes, *Res J Recent Sci,* 1:1–8.

Kurtz TW., Morris RC., Pershadsingh HA., (1989) The Zucker fatty rat as a genetic model of obesity and hypertension, *Hypertension,* 3:896–901.

Laemmli, UK., (1970) Cleavage of structural proteins during the assembly of the head of bacteriophage T4. *Nature,* 227: 680-685.

Lagnajdo R., Hardy M., (1967) Brain esterases during development, *Nature,* H N D. 214, 1207-1210.

Lakshmi P., Rajalakshmi P., (2011) Identification of phytocomponents and its biological activities of *Aloe vera* through the gas chromatography mass spectrometry, *Int Res J Pharm,* 2:247-249.

Lakshmi VP., Suseela M., Doss JP., (2015) Effect of cartap hydrochloride on lipid peroxidation in different tissues of albino rat, *Adv Res J Biol Sci Mol Tech,* 1:16-24.

Laloi C., Havaux M., (2015) Key players of singlet oxygen-induced cell death in plants, *Front Plant Sci,* 6:39.

Langmead L., Feakins RM., Goldthorpe S., (2004) Randomized, double-blind, placebo-controlled trial of oral *Aloe vera* gel for active ulcerative colitis. *Alim Pharmacol Ther* 19:739-747.

Lasram MM., Annabi AB., Elj N., El-Fazaa S., Gharbi N., (2009) Metabolic disorders of acute exposure to malathion in adult Wistar rats, *J Hazard Mater,* 163:1052-1055.

Lee KY., Weintraub ST., Yu BP., (2000) Isolation and identification of a phenolic antioxidant from *Aloe barbadensis*, *Free Radic Biol Med,* 28:261-265.

Lee YS., Ju HK., Kim YJ., Lim TG., Uddin MR., Kim YB., Baek JH., Kwon SW., Lee KW., Seo HS., Park SU., Yang TJ., (2013) Enhancement of anti-inflammatory activity of *Aloe vera* adventitious root extracts through the alteration of primary and secondary metabolites via salicylic acid elicitation, *PLoS One,* 8, e82479.

Lewis N., Williams TW., Chipman K., (2004) Functional analysis of xenobiotic response elements (XREs) in CYP 1A of the European Flounder (*Platichthys flesus*), *Mar Environ Res.*, 58:101-105.

Li X., Fang P., Mai J., Choi ET., Wang H., Yang XF., (2013) Targeting mitochondrial reactive oxygen species as novel therapy for inflammatory diseases and cancers. *J Hematol Oncol,* 6:19.

Liao JW., Kang JJ., Jeng CR., Chang SK., Kuo MJ., Wang SC., Liu MR., Pang VF., (2006) Cartap-induced cytotoxicity in mouse C2C12 myoblast cell line and the roles of calcium ion and oxidative stress on the toxic effects, *Toxicology,* 219:73-84.

Liao JW., Kang JJ., Liu SH., Jeng CR., Cheng YW., Hu CM., Tsai SF., Wang SC., Pang VF., (2000) Effects of cartap on isolated mouse phrenic nerve diaphragm and its related mechanism, *Toxicol Sci,* 55:453-9.

Liao JW., Pang VF., Jeng CR., Chang SK., Hwang JS., Wang SC., (2003) Susceptibility to cartap-induced lethal effect and diaphragmatic injury via ocular exposure in rabbits, *Toxicol,* 5:139-48.

Liao Y., Lu X., Lu C., Li G., Jin Y., Tang H., (2008) Selection of agents for prevention of cisplatin-induced hepatotoxicity, *Pharmacol Res,* 57:125-131.

Lindsey K., Jäger A., Viljoen A., vanWyk BE., (2002) Cyclooxygenase inhibitory activity of Aloe species, *S Af J Bot,* 68:47–50.

Ling LT., Palanisamy UD., Cheng HM., (2010) Prooxidant/antioxidant ratio (ProAntidex) as a better index of net free radical scavenging potential, *Mole,* 15:7884-92.

Liu J., Head E., Gharib AM., Yuan W., Ingersoll RT., Hagen TM., Cotman CW., Ames BN., (2002) Memory loss in old rats is associated with brain mitochondrial decay and RNA/DNA oxidation: partial reversal by feeding acetyl-L-carnitine and/or R-alpha -lipoic acid, *Proc Natl Acad Sci U S A,* 99:2356–61.

Long R., Satoh MH., Martin BM., Kimura S., Gonzalez FJ., Pohl LR., (1988), Rat liver carboxylesterase cDNA cloning, sequencing and evidence for a multigene family, *Biochem Biophys Res Commun,* 156: 866.

Loots DT., Van Der Westhuizen FH., Botes L., (2007) *Aloe ferox* leaf gel phytochemical content, antioxidant capacity, and possible health benefits, *J Agric Food Chem,* 55:6891–6.

Lopez A., Tangil MSD., Vega-Orellana O., Ramirez AS., Rico M., (2013) Phenolic constituents, antioxidant and preliminary antimycoplasmic activities of leaf skin and flowers of *Aloe vera* (L.) Burm. f. (syn. *A. barbadensis* Mill.) from the Canary Islands (Spain), *Mole,* 18:4942-4954.

Lopez Z., Nunez-Jinez G., Avalos-Navarro G., Rivera G., Salazar-Flores J., Ramirez J.A., Ayil-Gutierrez B.A., Knauth P., (2017) Antioxidant and cytotoxicological effects of *Aloe vera* food supplements, *J Food Quality,* 2017(7636237), 10.

Lowry, O. H., Rosebrough, N. J., Farr, A. L., Randel, R. J., (1951) Protein measurement with Folin-phenol reagent, *J Biol Chem,* 193(1), 265-275.

Madrigal-Santillan E., Madrigal-Bujaidar E., Alvarez-Gonzalez I., Sumaya-Martinez MT., Gutierrez-Salinas J., Bautista M., Morales-Gonzalez A., Gonzalez-Rubio GLM., Aguilar-Faisal JL., Morales-Gonzalez JA., (2014) Review of natural products with hepatoprotective effects, *World J Gastroenterol,* 20:4787-804.

Madrigal-Santillan E., Madrigal-Bujaidar E., Alvarez-Gonzalez I., Sumaya-Martinez MT., Gutierrez-Salinas J., Bautista M., Morales-Gonzalez A., Gonzalez-Rubio G-LM., Aguilar-Faisal JL., Morales-Gonzalez JA., (2014) Review of natural products with hepatoprotective effects, *World J Gastroenterol,* 20:14787-804.

Maiuri MC., Zalckvar E., Kimchi A., Kroemer G., (2007) Self-eating and self-killing: crosstalk between autophagy and apoptosis, *Nat Rev Mol Cell Biol,* 8:741–52.

Manvitha K., Bidya B., (2014) *Aloe vera*: A wonder plant its history, cultivation and medicinal uses, *J Pharmacog Phytochem,* 2:85-88.

Marco M., Michele N., Fabrizio C., Ferdinando M., Salvatore VG., Sebastiano G., Gioacchino, C., (2015) Review of clinical pharmacology of *A. vera* L. in the treatment of psoriasis, *Phytother Res,* 29:648-655.

Marklund S., Marklund G., (1974) Involvement of the superoxide anion radical in the autoxidation of pyrogallol and a convenient assay for superoxide dismutase, *Eur J Biochem,* 47:469-474.

Martindale JL., Holbrook NJ., (2002) Cellular response to oxidative stress: signaling for suicide and survival, *J Cell Physiol,* 192:1–15.

Martínez-Romero D., Castillo S., Zapata PJ., Valero D., (2013) Is it possible to increase the aloin content of *Aloe vera* by the use of ultraviolet light? *J Agricul Food Chem*, 61:2165-2170.

Maugh II, Thomas H., (2010) *Study links pesticide to ADHD in children*, Los Angeles Times.

Mauro P., Renze B., Wouter W., (2006) Enzymes. In: *Tietz text book of clinical chemistry and molecular diagnostics*, Carl AB, Edward R, David EB. Elsevier, 604-616.

Mauro P., Renze B., Wouter W., (2006) Enzymes. In: *Tietz text book of clinical chemistry and molecular diagnostics,* Elsevier, 604-616.

MC., (2007) Malathion exposure and the incidence of cancer in the agricultural health study, *Am J Epidemiol*, 166:1023–34.

McGovern M., Theodore M., (1998) Botanical dermatology, *Int J Dermatol,* 37:321-334.

Meister A., (1994) Glutathione-ascorbic acid antioxidant system in animals, *J Biol Chem,* 269:9397–400.

Melov S., Schneider JA., Day BJ., Hinerfeld D., Coskun P., Mirra SS., Crapo JD., Wallace DC., (1998) A novel neurological phenotype in mice lacking mitochondrial manganese superoxide dismutase, *Nat Gene,* 18:159–63.

Michalek H., Meneguz A., Bisso GM., (1978) Influence of the experimental conditions on the electrophoretic separation of AChE isoenzymes of the rat brain. *7th International Congress of Pharmacology*, Paris, France, Abstract 214.

Miller GT., (2002) *Living in the Environment*, Belmont: Wadsworth/Thomson Learning. ISBN 9780534376970.

Miller GT., (2004) *Sustaining the Earth*, Pacific Grove, CA: Thompson Learning, Inc, 211–216.

Miller RA., Britigan BE., (1997) Role of oxidants in microbial pathophysiology, *Clin Microbiol Rev,* 10:1–18.

Mink PJ., Mandel JS., Lundin JI., Sceurman BK., (2011) Epidemiologic studies of glyphosate and non-cancer health outcomes: a review, *Regul Toxicol Pharmacol,* 61:172–84.

Mizuno N., Niwa T., Yotsumoto Y., Sugiyama Y., (2003) Impact of drug transporter studies on drug discovery and development, *Pharmacol Rev,* 55:425–461.

Mizuuchi Y., Shi SP., Wanibuchi K., Kojima A., Morita H., Noguchi H., Abe I., (2009) Novel type III polyketide synthases from *Aloe arborescens*, *FEBS J,* 276:2391-2401.

Moemeka AN., Hildebrandt AL., Radaskiewicz P., King TR., (1998), Shorn (shn): a new mutation causing hypotrichosis in the Norway rat, *J Heredity,* 89:257–60.

Mohammed MI., Hafez SE., (2000), Biochemical studies on the protein content of fat body and haemolymph of fourth larval instar of *Musca domestica* L.; adult haemolymph and gonads, both emerging from reciprocal crosses after treatment with an organophosphorus pesticide, *J Union Arab Biol Cairo,* 13:33-51.

Monosson E., (2005) Chemical mixtures: considering the evolution of toxicology and chemical assessment, *Environ Health Perspect,* 113:383-390.

Moore PD., Yedjou CG., Tchounwou PB., (2010) Malathion-induced oxidative stress, cytotoxicity, and genotoxicity in human liver carcinoma (HepG2) cells, *Environ Toxicol,* 25:221-6.

Mordes JP., Bortell R., Blankenhorn EP., Rossini AA., Greiner DL., (2004) Rat models of type 1 diabetes: genetics, environment, and autoimmunity, *ILAR Journal,* 45:278–91.

Morello-Frosch R., Zuk M., Jerrett M., Shamasunder B., Amy DK., (2011) Understanding the Cumulative Impacts of Inequalities in Environmental Health: Implications for Policy, *Health Aff*, 30:879–87.

Moriyama M., Moriyama H., Uda J., Kubo H., Nakajima Y., Goto A., Akaki J., Yoshida I., Matsuoka N., Hayakawa T., (2016) Beneficial effects of the genus *Aloe* on wound healing, cell proliferation, and differentiation of epidermal keratinocytes, *PLoS One,* 11: e0164799.

Mueller S., Riedel HD., Stremmel W., (1997) Direct evidence for catalase as the predominant H_2O_2-removing enzyme in human erythrocytes, *Blood,* 90:4973–8.

Muller F., (2000) The nature and mechanism of superoxide production by the electron transport chain: Its relevance to aging, *J Am Aging Assoc,* 23:227–53.

Muller FL., Lustgarten MS., Jang Y., Richardson A., Van Remmen H., (2007) Trends in oxidative aging theories, *Free Radic Biol Med,* 43:477–503.

Multhoff G., Molls M., Radons J., (2012) Chronic inflammation in cancer development, *Front Immunol,* 2:98.

Murphy PJ., (2001), Xenobiotic metabolism: a look from the past to the future, *Drug Meta Disp,* 29:779–780.

Mustacich D., Powis G., (2000) Thioredoxin reductase, *Biochem J,* 346:1–8.

Mwanga H., Dalvie M., Singh T., Channa K., Jeebhay M., (2016) Relationship between pesticide metabolites, cytokine patterns, and asthma-related outcomes in rural women workers, *Int J Environ Res Public Health,* 13:957.

Nachiappan V., Muthukumar K., (2010) Cadmium-induced oxidative stress in *Saccharomyces cerevisiae*, *Ind J Biochem Biophy,* 47:383-387.

Ncibi S., Othman MB., Akacha A., Krifi MN., Zourgi L., (2008) *Opuntia ficusindica* extract protects against chlorpyrifos-induced damage on mice liver, *Food Chem Toxicol,* 46:797-802.

Nelson DR., Koymans L., Kamataki T., Stegeman JJ., Feyereisen R., Waxman DJ., Waterman MR., Gotoh O., Coon MJ., Estabrook RW., Gunsalus IC., Nebert DW., (1996) P450 superfamily: update on new sequences, gene mapping, accession numbers and nomenclature, *Pharmacogene,* 6:1-42.

Ni Y., Turner D., Yates KM., Tizard I., (2004) Isolation and characterization of structural components of *Aloe vera* L. leaf pulp. *Int Immunopharmacol,* 4:1745-1755.

Niehaus WGJr., Samuelsson B., (1968) Formation of malonaldehyde from phospholipid arachidonate during microsomal lipid peroxidation, *Eur J Biochem,* 6:126-130.

Nielsen SS., (2017) In: *Food Analysis,* fifth ed. Springer, New York, 259–260.

Nordberg J., Arnér ES., (2001) Reactive oxygen species, antioxidants, and the mammalian thioredoxin system, *Free Rad Biol Med,* 31:1287–312.

Nozik-Grayck E., Suliman HB., Piantadosi CA., (2005) Extracellular superoxide dismutase, *Int J Biochem Cell Biol,* 37:2466–71.

Obrien RD., (1967) *Insecticides.* Academic Press, New York.

Omura T., Sato R., (1962). A new cytochrome in liver microsomes. *J Biol Chem,* 237:375–1376.

Ordonez AAL., Gomez JD., Vattuone MA., (2006) Antioxidant activities of *Sechium edule* (Jacq.) Swart extracts, *Food Chem,* 97:452-458.

Organophosphates: Background, Pathophysiology, *Epidemiology.* 2016.

Oyedemi SO., Bradley G., Afolayan AJ., (2010) In vitro and In vivo antioxidant activities of aqueous extract of *Strychonos henningsii* Gilg, *Afr J Pharm Pharmacol,* 4:70–8.

Ozben T., (2007) Oxidative stress and apoptosis: impact on cancer therapy, *J Pharm Sci,* 96:2181–96.

Ozden S., Alpertunga B., (2010) Effects of methiocarb on lipid peroxidation and glutathione level in rat tissues, *Drug Chem Toxicol,* 33:50-54.

Ozgen M., Schreerens JC., Reese RN., Miller RA., (2010) Total phenolic, anthocyanidin contents and antioxidant capacity of selected elderberry (*Sambucus canadensis* L.) accession, *Pharmacog Mag,* 6:198–203.

Padma VV., Sowmya P., Felix TA., Baskaran R., Poornima P., (2011) Protective effect of gallic acid against lindane induced toxicity in experimental rats, *Food Chem Toxicol,* 49:991-998.

Pagalia DE., Valentine WN., (1967) Studies on the quantitative and qualitative characterization of erythrocyte glutathione peroxidase, *J Lab Clin Med,* 70:158-69.

Pal S., Sahrawat A., Prakash D., (2013) *Aloe vera*: Composition, processing and medicinal properties, *Curr Disc,* 2:106-122.

Panche A., Diwan A., Chandra S., (2016) Flavonoids: an overview, *J Nutr Sci,* 5.

Park MY., Kwon HJ., Sung MK., (2009) Evaluation of aloin and aloe-emodin as anti-inflammatory agents in aloe by using murine macrophages, *Biosci Biotech Biochem,* 73:828-832.

Patel DK., Patel K. Dhanabal SP., (2012) Phytochemical standardization of *Aloe vera* extract by HPTLC techniques, *J Acute Disease,* 1:47-50.

Patel RP., Cornwell T., Darley-Usmar VM., (1999) The biochemistry of nitric oxide and peroxynitrite: implications for mitochondrial function. In Packer L, Cadenas E (eds.). *Understanding the process of aging: the roles of mitochondria, free radicals, and antioxidant*s. New York, NY: Marcel Dekker, 39-56.

Peter JV., Sudarsan TI., Moran JL., (2014) Clinical features of organophosphate poisoning: A review of different classification systems and approaches. *Ind J Crit Care Med,* 18(11):735–745.

Philip L., Iyengar L., Venkobacher L., (2000) Site of interaction of copper on *Bacillus polymyxa*, *Water Air Soil Poll,* 119:1121.

Pineiro-Carrero VM., Pineiro EO., (2004) Liver, *Pediatrics,* 113:1097-106.

Popovska-Gorevski M., Dubocovich ML., Rajnarayanan RV., (2017) Carbamate insecticides target human melatonin receptors, *Chem Res Toxicol,* 30:574–582.

Pugh N., Ross SA., ElSohly MA., Pasco DS., (2001) Characterization of aloeride, a new high-molecular-weight polysaccharide from *Aloe vera* with potent immunostimulatory activity, *J Agri Food Chem,* 49:1030-1034.

Rada B., Leto TL., (2008) Oxidative innate immune defenses by Nox/Duox family NADPH oxidases, *Contrib Microbiol,* 15:164–87.

Radha MH., Laxmipriya NP., (2014) Evaluation of biological properties and clinical effectiveness of *Aloe vera*: A systematic review, *J Tradit Complement Med,* 5:21-6.

Rai DK., Rai PK., Gupta A., Watal G., Sharma B., (2009) Cartap and carbofuran induced alterations in serum lipid profile of Wistar rats, *Ind J Clinic Biochem,* 24:198-201.

Rai DK., Sharma B., (2007) Carbofuran-induced oxidative stress in mammalian brain, *Mol Biotech,* 37:66-71.

Rai DK., Sharma RK., Rai PK., Watal G., Sharma B., (2011) Role of aqueous extract of *Cynodon dactylon* in prevention of carbofuran-induced oxidative stress and acetylcholinesterase inhibition in rat brain, *Cell Mol Biol* (Noisy-le-grand), 57:135-142.

Rajasekaran, S., Ravi K., Sivagnanam K., Subramanian S., (2006) Beneficial effects of *Aloe vera* leaf gel extract on lipid profile status in rats with streptozotocin diabetes, *Clin Exp Pharmacol Physiol*, 33:232-237.

Rajendran UM., Kathirvel E., Anand N., (2007) Desiccation-induced changes in antioxidant enzymes, fatty acids, and amino acids in the cyanobacterium *Tolypothrix scytonemoides. World J Micro Biotech* 23, 251–257.

Ramachandra CT., Rao PR., (2008) Processing of *Aloe vera* leaf gel: A review, *Am J Ag Biol Sci,* 3:502-510.

Ramsey MR., Sharpless NE., (2006) ROS as a tumour suppressor? *Nat Cell Biol,* 8:1213–5.

Rao GV., Rupela OP., Rao VR., Reddy YV., (2007) Role of biopesticides in crop protection: present status and future prospects, *Ind J Plant Prot,* 35:1–9.

Raphael E., (2012) Phytochemical constituents of some leaves extract of *Aloe vera* and *Azadirachta indica* plant species, *Glob Adv Res J Environ Sci Toxicol,* 1:14–17.

Ray DE., (1991) Pesticides derived from plants and other organisms, *Handbook of Pesticide Toxicology,* 585-636.

Reitman S., Frankel S., (1957) A colorimetric method for the determination of serum glutamic oxaloacetic and glutamic pyruvic transaminases, *Am J Clin Pathol,* 28:56-63.

Renschler MF., (2004) The emerging role of reactive oxygen species in cancer therapy, *Eur J Cancer,* 40:1934–40.

Research Animal Models. *CRiver.com.* Charles River Laboratories, archived from the original on 24 May 2013, Retrieved 5 August 2012, https://www.criver.com/products-services/find-model.

Reynolds, T., Dweck AC., (1999) *Aloe vera* gel leaf: a review update, *J Ethnopharmacol,* 68:3-37.

Rhee SG., Chae HZ., Kim K., (2005) Peroxiredoxins: a historical overview and speculative preview of novel mechanisms and emerging concepts in cell signaling, *Free Rad Biol Med,* 38:1543–52.

Ribeiro-Santos R., Carvalho-Costa D., Cavaleiro C., Costa HS., Albuquerque TG., Castilho MC., Castilho MC., Ramos F., Melo NR., Sanches-Silva AA., (2015) novel insight on an ancient aromatic plant: The rosemary (*Rosmarinus officinalis* L.), *Trends Food Sci Tech,* 45:355-368.

Rodríguez ER., Martín JD., Romero CD., (2010) *Aloe vera* as a functional ingredient in foods, *Crit Rev Food Sci Nut,* 50:305-326.

Rodriguez F., Baldassarre H., Simonetti J., Aste F., Ruttle JL., (1988) Cervical versus intrauterine insemination of ewes using fresh or frozen semen diluted with *Aloe vera* gel, *Theriogenol,* 30:843-54.

Ross D., (1988) Glutathione, free radicals and chemotherapeutic agents. Mechanisms of free-radical induced toxicity and glutathione-dependent protection, *Pharmacol Therap,* 37:231-249.

Rothlein J., Rohlman D., Lasarev M., (2006) Organophosphate pesticide exposure and neurobehavioral performance in agricultural and non-agricultural Hispanic workers, *Environ Health Perspect,* 114:691-696.

Saafi EB., Louedi M., Elfeki A., Zakhama A., Najjar MF., Hammamia M., Achour L., (2011) Protective effect of date palm fruit extract (*Phoenix dactylifera* L.) on dimethoate induced-oxidative stress in rat liver, *Exp Tox Path*, 63:433-441.

Saccu, D., Bogoni P., Procida G., (2001) *Aloe* exudate: characterization by reversed phase HPLC and headspace GC-MS, *J Agri Food Chem,* 49:4526-4530.

Sadek M., Samaan H., Garawany El-A., Garawany A.E., (1989) The *in vivo* and *in vitro* inhibition of serum aminotransferases by acetylcholinesterases by anticholinesterase insecticides in rats, *Egypt Pharmaceut Sci,* 30:437-44.

Sahu PK., Giri DD., Singh R., Pandey P., Gupta S., Shrivastava AK., Kumar A., Pandey KD., (2013) Therapeutic and medicinal uses of *Aloe vera*: A review, *Pharmacol Pharm,* 4:599-610.

Salehi B., Albayrak S., Antolak H., Kregiel D., Pawlikowska E., Sharifi-Rad M., Yadav Uprety., Fokou PVT., Yousef Z., Zakaria ZA., Elena Maria Varoni EM., Sharopov F., Martins N., Iriti M., Sharifi-Rad J., (2018) *Aloe* genus plants: from farm to food applications and phytopharmacotherapy, *Int J Mol Sci,* 19:2843.

Sanborn M., Kerr KJ., Sanin LH., Cole DC., Bassil KL., Vakil C., (2007) Non-cancer health effects of pesticides: systematic review and implications for family doctors, *Can Fam Physician,* 53:1712–20.

Sanchez-Machad DI., Lopez-Cervantes J., Sendon R., Sanches-Silva A., (2017) *Aloe vera*: Ancient knowledge with new frontiers, *Trends Food Sci Tech,* 61:94-102.

Sawant O., Kadam VJ., Ghosh R., (2009) In vitro free radical scavenging and antioxidant activity of Adiantum lunulatum, *J Herb Med Tox,* 3:39–44.

Scherz-Shouval R., Elazar Z., (2007) ROS, mitochondria and the regulation of autophagy, *Trends Cell Biol,* 17:422–7.

Schumacker PT., (2006) Reactive oxygen species in cancer cells: live by the sword, die by the sword, *Cancer Cell,* 10:175–6.

Sedlak J., Lindsay RH., (1968) Estimation of total, protein bound, and nonprotein sulfhydryl groups in tissue with Ellman's reagent, *Anal Biochem,* 25:192-205.

Selmi S., Rtibi K., Grami D., Sebai H., Marzouki L., (2018) Malathion, an organophosphate insecticide, provokes metabolic, histopathologic and molecular disorders in liver and kidney in prepubertal male mice, *Tox Rep,* 5:189-195.

Shakoori AR., Saleem M.A., (1989) Somemacromolecular abnormalities developed byinteraction of Malathion and permethrin andsubsequent refeeding in *Tribolium castaneum* larvae, *Arch Insect Biochem Physiol,* 11: 203-215.

Sharma R., Yang Y., Sharma A., Awasthi S., Awasthi YC., (2004) Antioxidant role of glutathione-S-transferases: protection against oxidant toxicity and regulation of stress-mediated apoptosis, *Antiox Red Sig,* 6:289–300.

Shelton RM., (1991) *Aloe vera*. Its chemical and therapeutic properties. *Int J Dermat,* 30:679-683.

Shrivastava A., Kuzontkoski PM., Groopman JE, Prasad A., (2011) Cannabidiol induces programmed cell death in breast cancer cells by coordinating the cross-talk between apoptosis and autophagy, *Mol Cancer Ther,* 10:161–72.

Siatka T., Kasparova M., (2010) Seasonal variation in total phenolic and flavonoid contents and DPPH scavenging activity of *Bellis perennis* L. *Flowers. Mole,* 15:9450-9461.

Sies H., (1993) Strategies of antioxidant defense, *Eur J Biochem/FEBS,* 215:213–9.

Sies H., (1997) Oxidative stress: oxidants and antioxidants, *Exp Physiol,* 82:291–5.

Sims GK., Sommers LE., (1986) Biodegradation of pyridine derivatives in soil suspensions, *Environ Tox Chem,* 5:503–509.

Singh PK. Tewari RK., (2003) Cadmium toxicity induced changes in plant water relations and oxidative metabolism of *Brassica juncea* L. plants, *J Environ Biol,* 24:107-112.

Sinha JK., Ghosh S., Swain U., Giridharan NV., Raghunath M., (2014) Increased macromolecular damage due to oxidative stress in the neocortex and hippocampus of WNIN/Ob, a novel rat model of premature aging, *Neurosci,* 269:256–64.

Sotaniemi EA., Pelkonen O., Arranto AJ., Tapanainen P., Rautio A., Pasanen M., (2002) Diabetes and elimination of antipyrine in man: an analysis of 298 patients classified by type of diabetes, age, sex, duration of disease and liver involvement, *Pharmacol Tox,* 90:155-160.

Stadtman ER., (1992) Protein oxidation and aging, *Sci,* 257:1220-4.

Takahashi A., Ohtani N., Yamakoshi K., Iida S., Tahara H., Nakayama K., Nakayama KI., Ide T., Saya H., Hara E., (2006) Mitogenic signalling and the p16INK4a-Rb pathway cooperate to enforce irreversible cellular senescence, *Nat Cell Biol,* 8:1291–7.

Takaya K., Ogawa Y., Isse N., Okazaki T., Satoh N., Masuzaki H., Mori K., Tamura N., Hosoda K., Nakao K., (1996) Molecular cloning of rat leptin receptor isoform complementary DNAs-identification of a missense mutation in Zucker fatty (fa/fa) rats, *Biochem Biophys Res Comm,* 225:75–83.

Tan, Z., Li F., Xing J., (2011) Separation and purification of *Aloe* anthraquinones using PEG/salt aqueous two-phase system, *Sep Sci Technol*, 46:1503-1510.

Taylor P., Radic Z., Hosea NA., (1995) Structural bases for the specificity of cholinesterase catalysis and inhibition, *Toxicol Lett,* 83:453-458.

Thornalley PJ., (1990) The glyoxalase system: new developments towards functional characterization of a metabolic pathway fundamental to biological life, *Biochem J,* 269:1-11.

Thummel KE., Wilkinson GR., (1998) *in vitro* and *in vivo* drug interactions involving human CYP3A, *Ann Rev Pharmacol Tox,* 38:389-430.

Tochhawng L., Deng S., Pervaiz S., Yap CT., (2013) Redox regulation of cancer cell migration and invasion, *Mito,* 13:6–53.

Toler SM., Noe D., Sharma A., (2006) Selective enhancement of cellular oxidative stress by chloroquine: implications for the treatment of glioblastoma multiforme, *Neurosurg Focus,* 21: E10.

Toth SJ., (1996) *Federal Pesticide Laws and Regulations*, North Carolina Cooperative Extension Service.

Trachootham D., Alexandre J., Huang P., (2009) Targeting cancer cells by ROS-mediated mechanisms: a radical therapeutic approach? *Nat Rev Drug Discov,* 8:579–91.

Trevor AJ., Gordon MA., Parker KK., Chan SL., (1978) Acetylcholinesterases, *Life Sci,* 23:1209-1220.

Tripathi A., John SA., Rao KP., Ramteke PW., (2013) Comparative study on phytochemical constituents, antimicrobial and plasmid curing property of *Aloe vera* and *Ocimum sanctum*, *J Ecophysiol Occup Health,* 4:1–6.

Turrens JF., (2003) Mitochondrial formation of reactive oxygen species, *J Physiol,* 552:335–44.

US EPA., (2007) Assessing Health Risks from Pesticides. *Pesticides: Topical & Chemical Fact Sheets.*

Valverde JM., Valero D., Martínez-Romero D., Guillen F., Castillo S., Serrano M., (2005) Novel edible coating based on *Aloe vera* gel to maintain table grape quality and safety, *J Agricul Food Chem,* 53:7807-7813.

Van Raamsdonk JM., Hekimi S., (2009) Deletion of the mitochondrial superoxide dismutase SOD-2 extends lifespan in *Caenorhabditis elegans*, *PLoS Genet,* 5:e1000361.

Vandenbergh JG., (2000), Use of house mice in biomedical research, *ILAR Journal,* 41:133–135.

Vazquez B., Avila G., Segura D., Escalante, B. (1996) Anti-inflammatory activity of extracts from *Aloe vera* gel, *J Ethnopharmacol,* 55:69-75.

Velioglu YS., Mazza G., Gao L., Oomah BD., (1998) Antioxidant activity and total phenolics in selected fruits, vegetables, and grain products, *J Agric Food Chem,* 46:4113–4117.

Vertuani S., Angusti A., Manfredini S., (2004) The antioxidants and pro-antioxidants network: an overview, *Cur Pharm Des,* 10:1677–94.

Vijaya VK., Oschowka JA., (1977) Brain acetylcholinesterase activity and multiplicity in the Bonnet monkey (*Macaca radiata*) and the Rhesus monkey (*Macaca mulatta*). *J Neurochem,* 28:1141-1143.

Vijayan VK., Brownson RH., (1974) Polyacrylamide gel electrophoresis of rat brain acetylcholinesterase: isoenzymes of normal rat brain. *J Neurochem,* 23:47-53.

Vincent HK., Taylor AG., (2006) Biomarkers and potential mechanisms of obesity-induced oxidant stress in humans, *Int J Obes,* 30:400-418.

Vinson JA., Al Kharrat H., Andreoli L., (2005) Effect of *Aloe vera* preparations on the human bioavailability of vitamins C and E, *Phytomed,* 12:760-765.

Wang JF., Wei DQ., Chou KC., (2008) Drug candidates from traditional Chinese *medicines, Curr Top Med Chem,* 8:1656-6165.

Werawatganon D., Linlawan S., Thanapirom K., Somanawat K., Klaikeaw N., Rerknimitr R., Siriviriyakul P., (2014) *Aloe vera* attenuated liver injury in mice with acetaminophen-induced hepatitis, *BMC Complement Alt Med,* 14, 229.

Werck-Reichhart D., Feyereisen R., (2000) Cytochromes P450: a success story, *Genome Biol,* 1:3003.

West AP., Shadel GS., Ghosh S., (2011) Mitochondria in innate immune responses, *Nat Rev Immunol,* 11(6):389-402.

Willson HR., (1996) Pesticide Regulations. In Radcliffe EB, Hutchison WD, Cancelado RE (eds.). *Radcliffe's IPM World Text book.* St. Paul: University of Minnesota, USA.

Wilson BW., Schenkejl L., Fryd M., (1971) Cholinergic receptor molecules and cholinesterase molecules at mouse skeletal muscle junctions, *Nature,* 234:207–209.

Wintola OA., Afolayan AJ., (2011) Phytochemical constituents and antioxidant activities of the whole leaf extract of *Aloe ferox* Mill, *Pharmaco Mag,* 7:325-333.

Wolfe K., Wu X., Liu RH., (2003) Antioxidant activity of apple peels, *J Ag Food Chem,* 51:609-614.

Wolt JD., Smith JK., Sims JK., (1996) Products and kinetics of cloramsulam-methyl aerobic soil metabolism, *J Agric Food Chem,* 44: 324–332.

Wood ZA., Schroder E., Harris RJ., Poole LB., (2003) Structure, mechanism and regulation of peroxiredoxins, *Trends Biochem Sci,* 28:32-40.

World Health Organization (WHO), (1993) *Biomarkers and risk assessment: concepts and principles environmental health criteria*, No. 155. Geneva, Switzerland.

Wozzniak A., Paduch R., (2012) *Aloe vera* extract activity on human corneal cells, *Pharmac Biol* (*Formerly Int J Pharmacog*), 50:147-158.

Wright C., Milne S., Leeson H., (2014) Sperm DNA damage caused by oxidative stress: modifiable clinical, lifestyle and nutritional factors in male infertility, *Rep Biomed Online,* 28:684–703.

Xie Z., Klionsky DJ., (2007) Autophagosome formation: core machinery and adaptations, *Nat Cell Biol,* 9:1102–9.

Yagi A., Kabash A., Mizuno K., Moustafa SM., Khalifa TI., Tsuji H., (2003) Radical scavenging glycoprotein inhibiting cyclooxygenase-2 and thromboxane A2 synthase from *Aloe vera* gel, *Plant Med*, 69:269-271.

Yang MC., McLean AJ., Rivory LP., Le Couteur DG., (2000) Hepatic disposition of neurotoxins and pesticides, *Pharmacol Toxicol,* 87:286-291.

Yeh GY., Eisenberg DM., Kaptchuk TJ., Phillips RS., (2003) Systematic review of herbs and dietary supplements for glycemic control in diabetes, *Diab Care,* 26:1277-1294.

Yoshimi N., Keita S., Fumihiko Y., Masahiro K., Toshihiko O., (2011) Extensive screening for herbal extracts with potent antioxidant properties, *J Clin Biochem Nutr,* 48:78–84.

Younes M., Sieger S., (1981) Mechanistic aspects of enhanced lipid peroxidation following glutathione depletion *in vivo*, *Chemico-Biol Interac,* 34:257-266.

Zidan N-H., Galal OA., (2012) Evaluation of hematological, cytogenetical and biochemical effets of malathion and spinosad on male Albino rats, *Am Eur J Tox Sci,* 4:118-130.

LIST OF PUBLICATIONS

RESEARCH ARTICLES (12)

1) Gupta VK., Siddiqi NJ., Sharma B., (2021) Ameliorative impact of *Aloe vera* on cartap mediated toxicity in the brain of Wistar rats, *Indian Journal of Clinical Biochemistry*, Published online.
2) Gupta VK., Kumar A., Pereira ML., Siddiqi NJ., Sharma B., (2020) Anti-Inflammatory and antioxidative potential of *Aloe vera* on the cartap and malathion mediated toxicity in Wistar rats, *International Journal of Environmental Research and Public Health*, 17(14):5177.
3) Gupta VK., Ojha AK., Siddiqi NJ., Sharma B., (2019) Hepatoprotective Effect of *Aloe vera* against cartap and malathion induced toxicity in Wistar rats, *Journal of Cellular Physiology,* 234(10):18329-18343.
4) Jaiswal SK., Gupta VK., Siddiqi NJ., Sharma B., (2017) Curcumin mediated attenuation of carbofuran induced toxicity in the heart of Wistar rats, *Cellular and Molecular Biology* (Noisy-le-grand), 63(6):12-17.
5) Gupta VK., Kumar A., Yadav SH., Pandey R., Sharma B., (2017) Acetylcholinesterase as a biomarker of arsenic induced cardiotoxicity in mammals, *Science International,* 5(4):142-149.

6) Singh A., Gupta VK., Siddiqi NJ., Tiwari S., Gopesh A., Sharma B., (2017) Time course studies on impact of low temperature exposure on the levels of protein and enzymes in fifth instar larvae of eri silkworm, *Philosamia ricini* (Lepidoptera: satuniidae), *Biochemistry and Analytical Biochemistry,* 6:321.
7) Jaiswal SK., Gupta VK., Ansari MD., Siddiqi NJ., Sharma B., (2017) Vitamin C acts as a hepatoprotectant in carbofuran treated rat liver slices *in vitro*, *Toxicology Reports,* 4:265-273.
8) Jaiswal SK., Sharma A., Gupta VK., Singh RK., Sharma B., (2016). Curcumin mediated attenuation of carbofuran induced oxidative stress in rat brain, *Biochemistry Research International,* 2016:7637931.
9) GuptaVK., Pathak A., Siddiqi NJ., Sharma B., (2016) Carbofuran modulates functions of acetylcholinesterase from rat brain *in vitro*, *Advances in Biology,* 2016:367520.
10) Gupta VK., Kumar A., Siddiqi NJ., Sharma B., (2015) Rat brain acetyl cholinesterase as a biomarker of cadmium induced neurotoxicity. *Open Access Journal of Toxicology,* 1(1):555553.
11) Jaiswal SK., Gupta VK., Siddiqi NJ., Pandey RS., Sharma B., (2015) Hepatoprotective effect of *citrus limon* fruit extract against carbofuran induced toxicity in Wistar rats, *Chinese Journal of Biology,* 2015:686071.
12) Gupta VK., Pal R., Siddiqi NJ., Sharma B., (2015) Acetylcholinesterase from human erythrocytes as a surrogate biomarker of lead induced neurotoxicity, *Enzyme Research,* 2015:370705.

REVIEW ARTICLES (10)

1) Gupta VK., Yarla NS., Pereira ML., Siddiqui NJ., Sharma B., (2020) Recent advances in ethnopharmacological and toxicological properties of bioactive compounds from *Aloe barbadensis* (Miller), *Aloe vera*, *Current Bioactive Compounds,* 16:1.

2) Singh N., Kumar A., Gupta VK., Sharma B., (2018) Biochemical and molecular bases of lead-induced toxicity in mammalian systems and possible mitigations, *Chemical Research in Toxicology,* 31(10):1009-1021.
3) Gupta VK., Sharma B., (2018) Phytochemicals from indian traditional plants as natural potential therapeutic agents against adenosquamous lungs, *Pharmacologia,* 9(1):39-45.
4) Singh N., Gupta VK., Kumar A., Sharma B., (2017) Synergistic effects of heavy metals and pesticides in living systems, *Frontiers in Chemistry,* 5:70.
5) Gupta VK., Sharma B., (2017) Role of phytochemicals in neurotrophins mediated regulation of Alzheimer's disease, *International Journal of Complementary and Alternative Medicine,* 7(4):00231.
6) Gupta VK., Singh R., Sharma B., (2017) Phytochemicals mediated signalling pathways and their implications in cancer chemotherapy: challenges and opportunities in phytochemicals based drug development: a review, *Biochemical Compounds,* 5:2.
7) Gupta VK., Sharma B., (2017) Forensic applications of indian traditional toxic plants and their constituents, *Forensic Research and Criminology International Journal,* 4(1):00101.
8) Gupta VK., Sharma B., (2016) Alleviation of acute poisoning of organophosphates in humans, *Anatomy Physiology and Biochemistry International Journal,* 1(2):555558.
9) Gupta VK., Sharma B., (2016) Modulations of mammalian brain function by antidepressant drugs: role of some phytochemicals as prospective antidepressant, *Evidence Based Medicine and Practice,* 2.
10) Gupta VK., Singh S., Agrawal A., Siddiqi NJ., Sharma B., (2015) Phytochemicals mediated remediation of neurotoxicity induced by heavy metals, *Biochemistry Research International,* 2015:534769.

BOOK CHAPTERS (03)

1) Pandey R., Tiwari P., Prakash V., Gupta VK., Sharma B., (2019) Free radicals and antioxidant defense in asthma. in *oxidative stress and antioxidant defense: biomedical value in health and diseases*, Nova Science Publishers, Inc.
2) Kumar A., Singh N., Pandey R., Gupta VK., Sharma B., (2018) Biochemical and molecular targets of heavy metals and their actions, in: Rai M., Ingle A., Medici S., (eds.) *Biomedical Applications of Metals.* Springer, Cham.
3) Gupta VK., Sharma B., (2015) Environmental hazards due to xenobiotics contamination: growing risk to human health and possible remedies, *Green Social Work: Environmental Protection*, Petagon Press, ISBN 978-81-8274-852-1.

ABSTRACTS/PAPER PRESENTATIONS IN NATIONAL/INTERNATIONAL CONFERENCES/ SYMPOSIA/MEETINGS/WORKSHOPS (20)

1) Gupta VK., Sharma B. Protective effect of *Aloe vera* on malathion and cartap induced toxicity in liver of wistar rats. *22nd Agricultural Scientists and Farmers Congress on Post harvest technology and management for empowering the rural society and employment generation* on 22-23rd February 2020 organized by Bioved Research Institute of Agriculture Technology and Sciences, Prayagraj.
2) Gupta VK., Sharma B. Renoprotective effect of *Aloe vera* extract against the toxicity of cartap and malathion in Wistar rats, *International Conference on Translational Biotechnology*, Biosangam 2020, February 21-23, 2020 organized by Department of Biotechnology, MNNIT, Prayagraj-211004, Uttar Pradesh, India at MNNIT, Prayagraj.

3) Gupta VK., Sharma B. *Protective effect of Aloe vera on cartap and malathion induced toxicity in the liver of Wistar rats*. Organized by The Academy of Environmental Biology & NBFGR- 29-30 October 2019 at Lucknow, UP-India.
4) Gupta VK., Sharma B. *Parthenium hysterophorus* leaf extracts contains inhibitory potential of HIV-RT activity. *4th Asia Pacific AIDS & Co-infections Conference* (*APACC*) is scheduled to take place on 27-29 June 2019 at Hong Kong, Taiwan.
5) Gupta VK. *Successfully completed the certificate of participation at the RTG Big Data Research Allahabad Summer School-by Allahabad University and the Interdisciplinary center for Scientific Computing* (*IWR*) of Heidelberg University from April 1st to April 5th 2019.
6) Gupta VK., Sharma B. Protection of *Aloe vera* leaf on cartap and malathion induced hepatotoxicity in the liver of wistar rats. National *Conference on Interdisciplinary Advancements in Biochemistry* (*NCIAB-2019*), 28th-29th March, 2019 at Department of Biochemistry, Dr. RMLAU, Ayodhya (U.P., India).
7) Gupta VK., Sharma B. Nephroprotective effect of *Aloe vera* leaf against cartap induced alterations in Wistar rats. *3rd ACSTM and 6th ACSE Annual Conference in Deira Dubai*, UAE during February 12-14, 2019.
8) Gupta VK., Sharma B. Acetylcholinesterase as a biomarker of arsenic induced cardiotoxicity in mammals. *Proceedings of 106th Indian Science Congress*, 3-7 January 2019 at Lovely Professional University, Phagwara, Jalandhar.
9) Jaiswal J., Singh R., Gupta VK., Kumar A., Sharma B. *Carissa carandas* and *Nelumbo nucifera* as Antioxidants and Anti-HIV Agents. *National Conference on Expanding Horizons of Biochemistry*, December 1, 2018 at Department of Biochemistry, University of Allahabad, U.P. India-211002.
10) Gupta VK., Kumar A., Siddiqi NJ., Sharma B. *Aloe vera* leaf extract ameliorates cartap induced toxicity in kidney of Wistar rats. *National Conference on Expanding Horizons of Biochemistry*, December 1,

2018 at Department of Biochemistry, University of Allahabad, U.P. India-211002.

11) Gupta VK., Sharma B. Extract of *Parthenium hysterophorus* leaves contains anti-HIV-RT potential. *4th Innovative Approaches for Identification of Antiviral Agents Summer School*, September 24-28, 2018 in Santa Margerita Di Pula, Sardinia, Italy.
12) Sharma B., Jaiswal SK., Gupta VK. Inhibition of mammalian acetylcholinesterase by carbofuran and protection by vitamin C and *Citrus Limon* fruit extract. Annual Congress on Medicinal Chemistry, Pharmacology and Toxicology. July 30-31, 2018 Amsterdam, Netherlands, *Journal of Organic and Inorganic Chemistry.* 2018, 4.
13) Gupta VK., Sharma B. Aqueous extract of *Aloe vera* leaf may ameliorate cartap mediated oxidative stress induced alterations in kidney of Wistar rats. *National Symposium on Biodiversity and Natural Resources for the Sustainable Development & 37th Annual Session of Academy of Environmental Biology* at CCS University, Meerut. From 24-26 November 2017.
14) Gupta VK., Jaiswal SK., Sharma B. Curcumin mediated protection from carbofuran induced toxicity in rat heart. *International Conference on Biological Sciences and Bio-Statistics* (*ICBS-2017*), at UPRTOU, Allahabad from March 9-10, 2017.
15) Gupta VK., Kumar A., Shweta., Sharma B. Methanolic extract of *C. Limon* leaf may protect DNA from damage. *86th Annual Session of NASI & the Symposium held at UCOST*, Dehradun from December 2-4, 2016.
16) Jaiswal SK., Gupta VK., Sharma B. Carbofuran induced toxicity in wistar rat liver may be ameliorated by pretreatment with *Citrus Limon* fruit extract. *International Conference on Translational Biotechnology*, Biosangam 2016, February 4-6, 2016 organized by Department of Biotechnology MNNIT Allahabad-211004, Uttar Pradesh, India at MNNIT, Allahabad.
17) Gupta VK. *National Workshop on Aquatic Resource Sustainability & Cutting-Edge Technologies Satellite Symposium on Emerging Trends in Skill Development*. April 10th -16, 2015, Parasitology Laboratory,

Department of Zoology, University of Allahabad, Allahabad-211002, U. P.., India.

18) Gupta VK. *Author's Workshop* Jointly Organized by Springer & University of Allahabad. 2015 at University of Allahabad, Allahabad.
19) GuptaVK., Pathak A., Siddiqi NJ., Sharma B. *in vitro* Modulation of rat brain acetylcholinesterase by carbofuran. *85th Annual Session of NASI & the Symposium on 'Marine and Fresh Water Ecosystems for National Development'* December 6-8, 2015 organized by KIIT University, Bhubaneswar - 751024, India at KIITS University, Bhubaneswar.
20) Gupta VK., Pal R., Siddiqi NJ., Sharma B. Lead mediated inhibition of human acetylcholinesterase. *International Conference on Emerging Challenges in Biotechnology, Human Health and Environment (RABAEB-2015) & 9th Annual Convention of the Association of Biotechnology and Pharmacy (ABAP) and Special Satellite Meeting on Molecular Neurochemical Networking Complexes in Alzheimer's Disease*, December 14-16, 2015 Jointly organized by Department of Biotechnology KL University, Vaddeswaram, Association of Biotechnology and Pharmacy and Department of Biotechnology, Acharya Nagarjuna University, Guntur, at KL University, Andhra Pradesh, India.

AWARDS AND HONORS (06)

1) Outstanding contribution in reviewing for Toxicology Reports (Elsvier)-2017.
2) Ambassdor for Bentham Science Publishers-2018.
3) Senior Research Fellowship-Indian Council of Medical Research-New Delhi-2018.
4) European Society of Virology award to present research work in 4th Innovative Approaches for Antiviral Agents Summer School-2018, University of Cagliari, Italy.

5) Awarded with travel Grant to attend 3rd ACSTM and 6th ACSE Annual Conference in Deira Dubai, UAE during February 12-14, 2019. Not Availed.
6) Awarded with the Young Scientist Award, at 22nd Indian Agricultural Scientists & Farmers Congress, held during February, 22-23, 2020 at BRIATS Sringverpur, Prayagraj.

INDEX

B

C

D

E

F

G

H

I

K

L

M

N

O

P

Q

R

S

T

U

V

W

X

Z